# LAND DEGRADATION, DESERTIFICATION AND CLIMATE CHANGE

Although much is known about the processes and effects of land degradation and climate change, little is understood about the links between them. Less still is known about how these processes are likely to interact in different social-ecological systems around the world, or how societies might be able to adapt to this twin challenge. This book identifies key vulnerabilities to the combined effects of climate change and land degradation around the world. It identifies triple-win adaptations that can tackle both climate change and land degradation, whilst supporting biodiversity and ecosystem services.

The book discusses methods for monitoring effects of climate change and land degradation, and adaptations to these processes. It argues for better co-operation and knowledge exchange, so that the research, land user and policy communities can work together more effectively to tackle these challenges, harnessing the "wisdom of crowds" to assess vulnerability and adapt to climate change and land degradation, whilst protecting livelihoods and biodiversity.

**Mark S. Reed** is a Professor of Socio-Technical Innovation at Newcastle University, UK. He was a Professor of Interdisciplinary Environmental Research at Birmingham City University until 2016, and prior to that, he was Director of the Aberdeen Centre for Environmental Sustainability at the James Hutton Institute and University of Aberdeen.

**Lindsay C. Stringer** is Professor in Environment and Development at the Sustainability Research Institute, School of Earth and Environment, University of Leeds, UK. She was Director of the Sustainability Research Institute from 2011–2014.

# LAND DEGRADATION, DESERTIFICATION AND CLIMATE CHANGE

## Anticipating, assessing and adapting to future change

*Mark S. Reed and Lindsay C. Stringer*

LONDON AND NEW YORK

from Routledge

First published 2016
by Routledge
2 Park Square, Milton Park, Abingdon, Oxon OX14 4RN

and by Routledge
711 Third Avenue, New York, NY 10017

*Routledge is an imprint of the Taylor & Francis Group, an informa business*

© 2016 Mark S. Reed, Lindsay C. Stringer and United Nations
Convention to Combat Desertification

*British Library Cataloguing-in-Publication Data*
A catalogue record for this book is available from the British Library

*Library of Congress Cataloging in Publication Data*
Names: Reed, Mark S. | Stringer, Lindsay C., 1979-
Title: Land degradation, desertification, and climate change: anticipating,
assessing, and adapting to future change / Mark S. Reed and Lindsay C.
Stringer.
Description: London ; New York : Routledge, 2016. | Includes
bibliographical
references and index.
Identifiers: LCCN 2015042076| ISBN 9781849712705 (hbk) | ISBN
9781849712712
(pbk) | ISBN 9780203071151 (ebk)
Subjects: LCSH: Land degradation. | Desertification. | Climatic changes.
Classification: LCC S627.C58 R44 2016 | DDC 333.7–dc23
LC record available at http://lccn.loc.gov/2015042076


ISBN: 978-1-84971-270-5 (hbk)
ISBN: 978-1-84971-271-2 (pbk)
ISBN: 978-0-203-07115-1 (ebk)


Typeset in Bembo
by Cenveo Publisher Services

# CONTENTS

# ILLUSTRATIONS

## Boxes

## Figures

# ACRONYMS AND ABBREVIATIONS

| | |
|---|---|
| CBD | Convention on Biological Diversity |
| CBO | Community-Based Organization |
| CCAFS | The CGIAR Research Program on Climate Change, Agriculture and Food Security |
| CGIAR | The Consultative Group for International Agricultural Research |
| $CH_4$ | Methane |
| $CO_2$ | Carbon dioxide |
| COP | Conference of the Parties |
| CRIC | Committee for the Review of the Implementation of the Convention |
| CSO | Civil Society Organization |
| CST | Committee on Science and Technology (of the UNCCD) |
| DDP | Dryland Development Paradigm |
| DESIRE | The EU-funded Desertification and Remediation of Degraded Land project |
| DLDD | Desertification, Land Degradation and Drought |
| rDNA | Ribosomal deoxyribonucleic acid |
| FAO | UN Food and Agriculture Organization |
| FWCC | First World Climate Conference |
| GDP | Gross Domestic Product |
| GEF | Global Environment Facility |
| GHG | Greenhouse Gas |
| GIS | Geographical Information System |
| GLASOD | Global Assessment of Soil Degradation |
| HASHI | Soil conservation programme from the Swahili "Hifadhi Ardhi Shinyanga" |
| IE | Impact evaluation |

| | |
|---|---|
| IPBES | Intergovernmental Platform on Biodiversity and Ecosystem Services |
| IPCC | Intergovernmental Panel on Climate Change |
| ITPS | Intergovernmental Technical Panel on Soils |
| IYDD | International Year of Deserts and Desertification |
| JLG | Joint Liaison Group |
| KE | Knowledge Exchange |
| LADA | FAO's Land Degradation Assessment in Drylands programme |
| LDC | Least Developed Country |
| LDCF | Least Developed Countries Fund |
| LDN | Land Degradation Neutrality |
| LDRA | Land Degradation and Restoration Assessment |
| LIBS | Laser-induced breakdown spectroscopy |
| MA | Millennium Ecosystem Assessment |
| MEA | Multi-lateral Environmental Agreement |
| NAP | National Action Programme |
| NAPA | National Adaptation Plan of Action |
| NGO | Non-Governmental Organization |
| NIRS | Near infrared reflectance spectroscopy |
| $N_2O$ | Nitrous Oxide |
| OECD | Organization for Economic Cooperation and Development |
| PACD | Plan of Action to Combat Desertification |
| REGATTA | Regional Gateway for Technology Transfer and Action on Climate Change in Latin America and the Caribbean |
| SDGs | Sustainable Development Goals |
| SLM | Sustainable Land Management |
| SPI | Science-Policy Interface |
| TEV | Total Economic Valuation |
| UN | United Nations |
| UNCCD | United Nations Convention to Combat Desertification |
| UNCOD | United Nations Conference on Desertification |
| UNEP | United Nations Environment Programme |
| UNESCO | United Nations Education, Scientific and Cultural Organization |
| UNFCCC | United Nations Framework Convention on Climate Change |
| UN-INWEH | United Nations Institute for Water, Environment and Health |
| USSR | Union of Soviet Socialist Republics |
| VOC | Volatile Organic Compound |
| WHO | World Health Organization |
| WMO | World Meteorological Organization |
| WOCAT | World Overview of Conservation Approaches and Technologies |

# ACKNOWLEDGEMENTS

This book would not have been possible without the input and guidance from a number of different people from across the world. Our starting point was an Impulse Report commissioned by the United Nations Convention to Combat Desertification (UNCCD) via the 3rd UNCCD Scientific Conference Scientific Advisory Committee. We would like to thank the following people for their contributions to the Impulse Report:

Farshad Amiraslani (University of Tehran, Iran), Martial Bernoux (Institut de Recherche pour le Développement (IRD), France), Daya Bragante (United Nations Economic Commission for Africa (UNECA)), Mauro Centritto (National Research Council, Italy), Joris De Vente (Spanish National Research Council (CEBAS-CSIC), Spain), German Kust (Moscow State University, Russia), Renaud Lapeyre (Institut du Développement Durable et des Relations Internationales (IDDRI), France), Ali Mahamane (Université de Maradi, Niger), Jose A. Marengo (CEMADEN MCTI, Brazil), Graciela I. Metternicht (University of New South Wales, Australia), Ana Maria Murgida (Universidad de Buenos Aires, Argentina), Robert S. Nowak (University of Nevada, USA), Snezana Oljaca (University of Belgrade, Serbia), Juan Antonio Pascual Aguilar (Centro de Investigaciones sobre Desertificación (CIDE), Spain) and Mary Seely (Desert Research Foundation of Namibia, Namibia). We also thank Ioan Fazey (University of Dundee, UK), Rosi Neumann (Birmingham City University, UK) and Steven Vella (Birmingham City University, UK).

In addition, we would like to thank members of the 3rd UNCCD Scientific Conference Scientific Advisory Committee who provided feedback on the Impulse Report: Mariam Akhtar-Schuster (Project Management Agency PT-DLR, Germany), Miriam Diaz (National Experimental University Francisco de Miranda, Venezuela), Cristobal Felix Diaz Morejon (Ministry of Science, Technology and the Environment, Cuba), Patrice Djamen Nana (Africa Conservation Tillage

Network, Burkina Faso), Gordana Grujic (Oasis association, Republic of Serbia), Sahibzada Irfanullah Khan (Civil Secretariat in Peshawar, Pakistan), Ashot Khoetsyan (Yerevan State University, Armenia), Pietro Laureano (International Traditional Knowledge Institute, Italy), William A. Payne (University of Nevada, USA), Melanie Réquier-Desjardins (International Center for Advanced Mediterranean Agronomic Studies – Agronomic Mediterranean institute of Montpellier (CIHEAM – IAMM), France), Graham P. Von Maltitz (Council for Scientific and Industrial Research, South Africa) and Tao Wang (Chinese Academy of Sciences, China).

We are extremely grateful to Uche Okpara for help in identifying additional case study material for this book, and Madie Whittaker for all her assistance with editing and formatting.

# FOREWORD

The year 2015 witnessed two major events which will shape the international environment and development agendas for the next decade and beyond: the Post 2015 Sustainable Development Goals (SDGs) and the new Paris Agreement on climate change which will enter into force in 2020. These agreements set the stage for how we, as the international community, tackle the greatest challenges of our time.

The broad scope of the SDGs and the manifold implications of climate change demand a consideration of many important aspects and interlinkages. But, if we are to find sustainable solutions for building resilient societies that respect planetary boundaries, we must give special attention to a particular set of interlinkages: those between climate change and land.

Climate change and land degradation have an iterative relationship, driving or exacerbating one another through positive and negative feedback loops. Higher temperatures, changing precipitation patterns and more extreme weather fuel the erosion of fertile soils through wind and water. In turn, severe land degradation, especially in the world's drylands, reduces the provision of ecosystem services with devastating consequences for food production, human well-being and the climate. Land degradation releases large amounts of carbon which exacerbates climate change. Since the nineteenth century, more than half of the terrestrial carbon that was stored in soils and vegetation has been lost through land degradation processes. Today, roughly one-quarter of all anthropogenic greenhouse gas emissions come from land use activities.

Fortuitously, these feedback loops also hold the promise of a better future because they work in the opposite way if harnessed for the good. The solutions for reversing land degradation have vast potential to mitigate and adapt to climate change. But that potential has been harnessed only marginally so far. The land sector is unique when it comes to mitigating climate change. It allows us to both reduce greenhouse gas emissions and to sequester carbon in the soil and in vegetation. There is growing recognition that limiting global warming to below a 2°C rise this century can only be realized if the mitigation potential of terrestrial ecosystems is utilized much more comprehensively. The largely untapped mitigation potential of terrestrial ecosystems offers one powerful and inexpensive path to bridge this gap in two concurrent steps, which form the basis of the Land Degradation Neutrality (LDN) concept: to avoid the degradation of healthy ecosystems and to restore and rehabilitate the degraded ecosystems.

In fact, the only way many developing countries can act on climate change is through land-based activities. This also applies to land-based adaptation strategies. Adapting land use to changing climatic conditions and making wise use of ecosystem services are widely considered to be key in building resilient societies and agro-ecosystems. Moreover, adaptation strategies such as Sustainable Land Management (SLM) render considerable co-benefits to rural livelihoods and the preservation of biodiversity.

Clearly, the achievement of the SDGs and limiting global warming demand that we take full account of the interlinkages between climate change and land degradation. This requires integrated approaches that maximize the synergies between climate change and land degradation while realizing and optimising the associated trade-offs. To this end, sound science to inform policy making is vital.

We know a lot about climate change. We are well informed about land degradation. However, we know a lot less about the interplay between the two processes. This was one of the main conclusions of the UNCCD 3rd Scientific Conference which was held in Cancun, Mexico, in March 2015. This book, which at the time served as an "impulse report" in preparation of this conference, is an important first bridge to address this knowledge gap. We are delighted to note that this report not only enriched the conference, but has evolved into a more comprehensive scientific work.

As we embark on implementing the SDGs and new climate agenda, this book comes at the right point in time. By explaining the interconnections between climate change and land degradation in a scientifically sound yet easy to understand manner, we are confident that this book will contribute to the identification of sustainable solutions which help to address the combined effects of land degradation and climate change. Such approaches are key to fostering the closer cooperation between our conventions on both policy and implementation levels.

Christiana Figueres,
Executive Secretary of the United
Nations Framework Convention
on Climate Change

Monique Barbut,
Executive Secretary of the United
Nations Convention to Combat
Desertification

# 1

# INTRODUCTION

Climate change and land degradation are closely interlinked and are most acutely experienced by ecosystems and resource-dependent populations in regions affected by desertification and drought. It is essential to understand and address the dual challenges of climate change and land degradation if we are to meet targets such as the Sustainable Development Goals (SDGs), tackle poverty and address many of the most pressing environmental challenges of the twenty-first century.

Although much is known about the processes and effects of land degradation and climate change, less is understood about the links between these two challenges. Less still is known about how climate change and land degradation processes are currently interacting in different social-ecological systems around the world, or how they might interact under different scenarios in the future. However, there are major challenges associated with anticipating the combined effects of climate change and land degradation processes, given numerous inherent, and often contradictory, feedbacks. Climate change and land degradation processes operate differently in different ecosystems[1], and within the same ecosystems under different forms of land management. This leads to a range of effects on ecosystem processes[2], which in turn influence the provision of ecosystem services[3] to society. This may give rise to a number of potentially important but possibly unforeseen impacts on populations in regions affected by Desertification, Land Degradation and Drought (DLDD)[4]. Moreover, limited understanding of feedbacks among these processes restricts our capacity for anticipatory adaptation. There is an increasingly urgent need for research to elucidate these links, so that land users and policy-makers can respond in timely and effective ways.

In this book we look at how land users, the policy and research communities and other stakeholders can work together to better anticipate, assess and adapt to the combined effects of climate change and land degradation in regions affected by DLDD. We also consider some of the behavioural, governance and policy changes that may be needed to facilitate effective adaptation to future change at national and international scales. We consider all regions affected by DLDD but place special emphasis on drier areas, because these are often considered most vulnerable to DLDD.

Using the United Nations Convention to Combat Desertification (UNCCD) definition which encompasses arid, semi-arid and dry sub-humid parts of the world, drylands occupy around 41 per cent of the Earth's land area and are home to around a third of the world's population (MA, 2005). The proportion of drylands thought to be affected by land degradation in the form of desertification depends largely on the definition of dryland, as well as the assessment method used, with estimates of 10 per cent (Lepers *et al.*, 2005), 38 per cent (Mabbutt, 1984), 64 per cent (Dregne, 1983), and 71 per cent (Dregne and Chou, 1992). A key attempt to quantify land degradation was undertaken in the Millennium Ecosystem Assessment, which suggests a figure of 10–20 per cent of drylands are degraded with "medium certainty" (MA, 2005), with degradation severity and extent highest in Africa and Asia[5].

At the same time as the challenge of land degradation, climate change is leading to global changes in temperature, rainfall, sea level rise, increasing concentrations of carbon dioxide ($CO_2$) and other greenhouse gases in the atmosphere, and an increase in the incidence and severity of extreme weather events. A possible temperature increase of 1–3°C in drylands, if $CO_2$ concentrations were to reach 700 p.p.m. by 2050, would increase global potential evapotranspiration by around 75–225 mm per year. Climate models have predicted that up to 50 per cent of the Earth's surface will be experiencing regular drought by the end of the twenty-first century under a "business as usual" scenario, with drylands in northern Africa, Amazonia, the United States, southern Europe and western Eurasia likely to become drier, while higher latitudes of the northern hemisphere are likely to become wetter (Burke *et al.*, 2006; Seager *et al.*, 2007; D'Odorico *et al.*, 2013). In more temperate locations however, higher temperatures may lengthen growing seasons (Cantagallo *et al.*, 1997; Travasso *et al.*, 1999). Elevated concentrations of atmospheric $CO_2$ are likely to have a fertilizing effect on plants, boosting primary productivity, and would likely increase the efficiency with which plants use water to create biomass (Le Houérou, 1996; Chun *et al.*, 2013; Keenan *et al.*, 2013; Kaminski *et al.*, 2014). At the same time, these effects are likely to be offset by negative effects of elevated tropospheric ozone and the impacts of changing distributions of weeds, pests and diseases, as well as changes to the composition of vegetation communities. In the chapters that follow, we explore these changes in more depth.

In this book we take an interdisciplinary and integrated approach to climate change and land degradation, considering them as interlinked concepts that have biophysical and human drivers, impacts and responses. Although interactions between climate change and land degradation are likely to give rise to a number of new challenges, there may also be a number of overlaps, as well as scope for synergy, between the behaviours, governance models and policy instruments that may be needed to address these issues. By bringing together scientific and other forms of knowledge from around the world (including local, traditional, indigenous and lay knowledge, which we collectively term "locally held knowledge"), it may be possible to reduce the vulnerability of ecosystems and populations to these threats, in currently affected areas and beyond, and to build overall resilience. We therefore consider how an integrated approach to addressing land degradation and climate change can be developed, in order to harness synergies and multiple benefits, as well as outlining some of the potential ways forward.

## 1.1 Climate change and land degradation in regions affected by DLDD: key definitions

Before attempting to understand the nature of the interlinkages between climate change and desertification, land degradation and drought, it is important to begin by providing some clarity in the definitions we are using – not least because they are all terms that are used differently by different stakeholders and researchers working within different disciplines.

## 1.2 Climate change

Climate can be thought of as a statistical description of the weather, taking into account variables including temperature, wind speed and direction, and rainfall, over a long time period. The World Meteorological Organization (WMO) usually considers this long time period to span from more than 30 years up to thousands of millions of years. Often, we think of the climate as being the conditions we experience at the Earth's surface. However, climate is really a summary of the state of the broader climate system, which includes a range of complex interactions between the atmosphere (the blanket of gases surrounding the Earth), hydrosphere (the water components present on the Earth), the cryosphere (the frozen parts of the planet) and the biosphere (parts of the Earth where life is found). The broader climate system has its own internal dynamics but is also affected by external biophysical phenomena such as volcanic eruptions on Earth, changes to the sun including variation in solar activity and the intensity of light energy, as well as human-induced changes in the composition of the atmosphere. This results in direct or indirect changes in the Earth's climate through feedbacks that operate on different timescales (IPCC, 2007).

According to the Intergovernmental Panel on Climate Change (IPCC, 2001), the primary international scientific body providing advice to the United Nations on climate-related challenges, climate change refers to a variation in climate that persists over decades or longer, that is statistically significant in terms of its mean state or its variability. Other definitions attempt to attribute climate change either directly or indirectly to human activities such as deforestation and industrialization, which change the balance of gases in the global atmosphere (e.g. UNFCCC, 1992, Article 1). Making links to human activities is very important for political/decision-making reasons, especially if international action is to be taken to address climate change. It presents the issue as more than just a natural occurrence, implicating humans in the problem and legitimating the need for policy action to develop solutions.

## 1.3 Defining land degradation and desertification

Land degradation is a process that can happen in any climatic zone, not just in drylands. Land degradation in drylands is sometimes referred to as "desertification". UNCCD (1994) defines desertification as "land degradation in arid, semi-arid, and dry sub-humid areas resulting from various factors, including climatic variations and human activities". UNCCD (1994) defines land degradation as a

> reduction or loss, in arid, semi-arid, and dry sub-humid areas, of the biological or economic productivity and complexity of rain-fed cropland, irrigated cropland, or range, pasture, forest, and woodlands resulting from land uses or from a process or combination of processes, including processes arising from human activities and habitation patterns, such as: (i) soil erosion caused by wind and/or water; (ii) deterioration of the physical, chemical, and biological or economic properties of soil; and (iii) long-term loss of natural vegetation.

This definition of land degradation therefore refers to:

*   Decline in biological and/or economic resilience (i.e. the ability of a system to maintain the structure essential to support the basic system functions (such as biological habitat, biomass production, filtering, buffering, storage and transformation of nutrients, water retention) during times of stress or perturbation (Holling, 1986; Ballayan, 2000); or
*   Loss of adaptive capacity (the ability – often measured in the time it takes – for a system to regain the structure essential to support basic system functions after stress or perturbation of the land system (Kasperson *et al.*, 1995; IPCC, 2001).

These considerations emphasize the importance of maintaining basic ecosystem processes, functions and services that may or may not include human uses. This approach to defining land degradation conceptualizes land as including all elements of the biosphere at or below the Earth's surface, incorporating soil, terrain, surface hydrology, groundwater, plants and animals, human settlements and the physical evidence of past and present human activity. As such, any approach to tackling land degradation needs to consider how to mitigawte impacts on underpinning ecosystem processes and prevent critical thresholds in natural capital from being crossed, in addition to mitigating the consequent loss of ecosystem services. For this reason, Reed *et al.* (2015: 472) argue that mechanisms for tackling land degradation need to be "based on retaining critical levels of natural capital whilst basing livelihoods on a wider range of ecosystem services".

The role of human activities in causing land degradation is important. For example, reductions in, or losses of, productivity and resilience can stem from soil erosion caused by wind and/or water; a loss of quality or integrity of the physical, chemical and biological or economic properties of soil and a loss or change in natural vegetation.

Each of these changes is driven largely by human activities such as land use change, mining and habitation patterns (including urbanization). Erosion can have particularly important economic impacts on agricultural land, where the redistribution of soil within a field, the loss of soil from a field, the breakdown of soil structure, and the decline in organic matter and nutrients, can all result in a reduction of cultivable soil depth and a decline in soil fertility (Morgan, 2005).

Despite the scientific evidence that supports the role of human activities as a key driver of land degradation, some scientists also consider that climatic variation (particularly drought) and longer-term drying out, aridification or 'desiccation' (due to climate change) are important contributing factors that underpin land degradation, particularly because desiccation can cause reductions in productivity and vegetation loss. Indeed, because drylands are water-limited environments, it can mean that water degradation (in terms of quality and quantity) can have substantial effects on both ecosystem integrity and human well-being.

Some scientists have suggested that land degradation can only be determined in relation to the goals of the management system at the time of investigation (e.g. Turner and Benjamin, 1993), and in the context of a specific time frame, spatial scale, economy, environment and culture (Warren, 2002). This means that the same bio-physical environmental change (e.g. erosion) can create different problems and have different consequences in different contexts. Warren argued that if soil erosion is

> of no consequence to production at a larger spatial scale, it does not contribute to degradation in the wider context. If it has no impact on future production, it is not degradation in the longer term. A change in a component of the environment that cannot be accessed with present technology or finance, or is inconsequential to a present way of life, does not, per se, amount to degradation.
>
> *(2002: 449)*

As such, the extent and severity of land degradation being experienced may vary between land users with different management goals in different places at different times and in different socio-economic, environmental, cultural and technological contexts.

In summary, land degradation: (i) is a phenomenon caused by human activities and exacerbated by certain climate, geological and topographic characteristics; (ii) is characterized by changes in ecosystem processes and levels of natural capital that affect the flow of ecosystem services to society; (iii) causes an effectively permanent decrease in the capacity of the land system as managed to meet its user demands; and (iv) is a threat to the long-term biological and/or economic resilience and adaptive capacity of the ecosystem and the populations who depend upon it.

## 1.4 Types of land degradation

Land degradation and drought affect a wide range of ecosystems and populations around the world and are not phenomena that are restricted to hot, dry areas that experience desertification. For example, Iceland experiences a climate that is maritime cold-temperate to sub-Arctic, with annual rainfall of the range of 500–2000 mm. However, the country experiences severe land degradation. Large-scale loss of trees, forests and other vegetation due to human activities that date back to the time of the Vikings (more than 1,130 years ago) mean that vast proportions of the country's land area are severely degraded.

Land can be considered degraded if losses or damage occur to soil, water or biomass. Water (both on the surface and under the ground) may be degraded in terms of its quality and its quantity (Box 1.1). For example, changes in vegetation cover can increase the speed with which water reaches watercourses, making flows more "flashy", leading to an increase in flash flooding immediately after major rainfall

---

## BOX 1.1: WATER DEGRADATION AND ITS LINKS TO LAND MANAGEMENT AND CLIMATE CHANGE

Water quality issues are closely linked to land quality, even beyond the drylands. Sao Paolo, in Brazil, suffered a multi-year drought starting in 2013 – its worst drought since the 1930s. The quality of the little remaining water was low, partly due poor land management in the surrounding watershed. The removal of riparian forest buffers that normally captured pollutants and held the soil in place led to higher concentrations of pollutants and sediments in the little water that remained.

Climatic factors can also play a role in land and water degradation in wetter parts of the world. For example, much of Bangladesh is low lying and nearly 50 per cent of the country's population lives within 10 m of average sea level. Climate change is likely to increase the salinity of important rivers that are used as drinking water and irrigation sources, as a result of both sea level rise and projected increases in storm surges linked to cyclone activity. This is expected to have important impacts on both aquatic ecosystems (including mangroves) and human well-being. Not only are the changes in water quality anticipated to affect the productivity of many capture fisheries, but also, a recent report by the World Bank drew attention to the impacts of climate-induced increases in soil salinity on high yielding varieties of rice in the country. The report suggests a 15.6 per cent decline in rice yields in areas where soil salinity (measured in electrical conductivity) is expected to be greater than four decisiemens per meter by 2050. The human impact of this is forecast to encompass higher dependency ratios, higher incidences of poverty, and increased out-migration of working age adults. Rice prices are projected to increase as a consequence and infrastructure maintenance costs are likely to rise. The report suggests that the rising salinity of groundwater will increase the demand for investment and expenditure on paved road repairs by more than 200 per cent of current rates.

Based on material from: Dasgupta *et al.* (2014a, b, c).

events, and exacerbating low flows in the absence of rainfall. Lower vegetation cover and more extensive areas of bare ground also increase the erosive potential of rainfall, increasing sediment and nutrient inputs to stream water, reducing water quality. Unsustainable rates of groundwater extraction, particularly in areas with coastal aquifers, can lead to salt-water intrusion, making groundwater unsuitable for irrigation.

Degradation of biomass (both above and below ground) can also constitute land degradation, as seen when bush encroachment takes place. Biomass degradation often occurs in areas used for livestock grazing and is linked to changes in species and community composition. Overgrazing can lead to the degradation of plant cover and the increase of certain species that are less palatable (or in some cases poisonous) to the main species of livestock that use the area. This reduces the productive potential of the land.

Degradation of the soil component of land takes four main forms: (i) water erosion; (ii) wind erosion; (iii) chemical degradation; and (iv) physical degradation, all eventually leading to biological degradation. The first two forms of soil degradation result in the loss or removal of soils in one location and their deposition in another. Often erosion removes the most fertile layer of topsoil, creating a negative implication for productivity and resilience in the place where it is being eroded (though sometimes a benefit in the area in which it is deposited). Areas of the world where soils are shallow and the land is particularly sloping are at risk from this type of degradation. The Ethiopian highlands, which cover approximately 45 per cent of the country, are one location where water erosion is highly problematic, causing the development of huge gullies in some parts (Selassie and Amede, 2014). Figures in the literature vary, but some consider more than two million hectares to be so degraded that the land is no longer able to support cultivation (EHRS, 1984). Aside from the visible damage to the landscape, water erosion has an economic cost as well. Estimates suggest that 1.9 billion tons of soil are lost each year through erosion, costing Ethiopia 2 to 3 per cent of its gross domestic product (GDP) from agriculture. These losses result in lower productivity of cropland and livestock, and knock-on impacts for those who depend on the land to support their livelihoods. Costs are not restricted to national boundaries. Erosion from the highlands costs downstream countries such as Egypt and Sudan between $280 million and $480 million each year to clear the sediments (Selassie and Amede, 2014).

Wind erosion is also highly problematic for the world's soils. In many places, for example in Argentina's pampas and northern parts of China, the rate of erosion by wind exceeds that by water. Agriculture is a dominant land use in these areas and the dust eroded from these lands can be transported across vast areas, even to other continents. Perhaps the best-known example of wind erosion is found in the USA's Great Plains in the twentieth century. The 1930s' dust bowl years were a period in which intensive farming without soil conservation measures, combined with a drought that lasted for several years, led to massive tracts of bare soil, which, under windy conditions, created large dust clouds and removed much of the topsoil. Many families abandoned their farms, while some of the people who remained behind died as a result of 'dust pneumonia' linked to the inhalation of tiny particles of soil (Worster, 2004). Wind erosion and the dust clouds it creates are now understood to have an important influence in shaping both regional and global climate.

Chemical degradation can take place independently of climate and relates to nutrient and organic matter losses, salinization, pollution, acidification and alkalinization. It can also trigger land use change by causing vegetation to die back. While soil acidification is a natural process, it is intensified by agriculture, as the concentration of hydrogen ions in the soil increases. Low soil pH can cause elements such as aluminium to become soluble. This can retard the growth of plant roots, which acts to limit crops from accessing soil water and nutrients. Similarly, in very acidic soils, important plant nutrients may be present but unavailable, leading to poor productivity and decreased microbial activity. In rangeland areas this can result in poor pastures and reduced livestock productivity. Unlike erosion, chemical degradation is not always immediately visible and may take time for the effects to be realized. This is particularly so for the chemical degradation of groundwater.

Nutrient and organic matter losses can often be remedied through the use of Sustainable Land Management (SLM) practices, such as minimum tillage and conservation agriculture, including the use of green manure and cover crops, which cover the soil to reduce erosion while also increasing fertility by fixing nitrogen. However, pollution and acidification can affect areas that are far away from the original site of emission, making responsibilities for addressing those forms of degradation more difficult to attribute. Such diffuse pollution (e.g. linked to sulphurous emissions during the smelting of mineral ores) can cross national borders, depending on the wind speed and direction.

Physical soil degradation is used to describe processes of compaction (which displace air from the small gaps between soil particles causing the soil to increase in density), crusting (development of a platy surface due to biological or physical processes), subsidence (which results from clay and silt soils shrinking when they dry out) and water logging (often due to a high groundwater level, causing the roots of plants in the soil to be starved of oxygen).

Soil sealing is a problematic form of physical soil degradation and often occurs during the construction of infrastructure, when the soil is covered with materials like concrete, plastic, etc., which prevent water from percolating through the land surface. This type of degradation is particularly problematic in developed, urban areas. For example, in Europe, approximately nine per cent of the total land area is sealed with impermeable material, with countries such as Germany, Portugal, the Netherlands, Spain and Hungary having sealed more than 45 per cent of their urban soils (Scalenghe and Marsan, 2009). Sealing has important effects on the relationship between the land and the climate. Surface sealing alters the balance of energy transfers and temperature regulation processes, affecting soil biota. Sealed surfaces absorb the sun's energy during the day and cool slowly at night. This can cause sealed surfaces to be up to 20 degrees warmer than unsealed surfaces. Sealing also affects water flows and gas diffusion, affecting the ability of the soil to act as a carbon sink and to filter air pollutants. This is an important and often overlooked consideration as the world's urban population continues to grow, as it provides important feedbacks that affect the climate.

These examples demonstrate the various forms of land degradation and also the close links between land and climate. They highlight that land degradation can occur under a range of different climatic conditions and land cover contexts, making it a problem throughout the world. International political attention was, however, first drawn to the problems of land degradation in the drylands.

## 1.5 Land degradation in drylands

In general, drylands are distributed between latitudes of approximately 20 and 35 degrees. Their presence in these parts of the planet is largely attributable to the global climate system (UNCCD, 1994; MA, 2005). Hyper-arid areas cover approximately 8 per cent of the world's total land area, and are mostly found within the Sahara, Gobi and Arabian deserts. Hyper-arid areas are not included in the UNCCD's definition of drylands because very few people live in these environments. The ratio of annual precipitation to potential evapotranspiration in hyper-arid areas is less than 0.05. Potential evapotranspiration is the amount of moisture that, if it were available, could potentially be lost from a given land area through processes of evaporation and transpiration. Water in hyper-arid areas is very scarce. This restricts both human activity and the production of biomass (vegetation). As a result, few crops are grown in the hyper-arid parts of the world, aside from those produced in oases or under irrigation (UNCCD, 1994; MA, 2005).

Arid, semi-arid and dry sub-humid regions are affected by an extreme form of land degradation known as "desertification". Arid, semi-arid and dry sub-humid areas are estimated to cover around 41 per cent of the Earth's surface. In these parts of the world, the ratio of annual precipitation to potential evapotranspiration falls within the range 0.05–0.65, so there is slightly more water in these parts than in hyper-arid areas (UNCCD, 1994). While water availability still represents an important limiting factor for agricultural production, arid, semi-arid and dry sub-humid areas nevertheless provide a wide range of commodities to the rest of the world, including fruit and vegetables, spices, meat, cotton, tobacco, fishery products, forest products and rubber (see Box 1.2). These commodities provide the main source of income for more than a billion people worldwide. For example, during the period 2003–2005, Africa's drylands produced 333,000 tons of citrus fruit, with 154,000 tons coming from Sudan alone. Over the same time period, crops such as coffee formed 60 to 70 per cent of export earnings from Ethiopia, Rwanda and Uganda (UNCCD and CFC, 2009). Cotton provided between 30 and 40 per cent of national export earnings for Burkina Faso, Chad and Mali, while spices such as ginger, pepper, cloves and chilli contributed important earnings to economies in countries such as Djibouti and Tanzania.

Drought can occur in areas other than the drylands. The research literature recognizes a range of different types of drought. For example, meteorological droughts are periods when there is dryness that exceeds the norm for a particular

---

## BOX 1.2: DRYLAND CROPS AND COMMODITIES

Many key crops such as maize, beans, potatoes and lentils originated from the drylands and date back to ancient times. For example, lentil remains in the drylands have been found together with human remains, in settlements thought to be older than 13,000 BC. Lentils are thought to be one of the first domesticated crops in the Old World, while maize, beans, tomato and potatoes originated from the dryland areas of Mexico, Peru, Bolivia and Chile. These crops are some of the world's major staple foods today and make an important contribution to global food security.

Some of the dryland areas such as those in the Middle East, North America and West Asia are rich in oil and play a key role in supporting the world's petroleum-dependent activities. All of the world's top ten oil producers (Russia, Saudi Arabia, USA, China, Canada, Iraq, Iran, UAE, Kuwait and Mexico) have extensive dryland operations. As focus is shifting towards lower carbon energy sources that come from sources other than fossil fuels, interest has grown in the world's drylands as areas in which oil seeds such as soybean and sunflower may be grown. Sudan and Ethiopia already produce sesame seeds and oil, raising $100 million and $50 million of export earnings respectively. In countries such as India and Mali, governments and investors have been exploring biofuel crop *Jatropha curcas* as an alternative energy source for both transportation and rural electrification. *Jatropha* is considered particularly promising in drylands due to its drought resistance, its low water demand and its low nutrient requirements, making it possible to grow *Jatropha* in areas with poor soils. Village level plantations and processing activities offer scope to provide fuel for household consumption, as well as running motors that have sufficient power to pump water. However, more research is required to investigate the agronomic, ecological and economic aspects of dryland *Jatropha* production (Favretto *et al.*, 2014). Drylands also have an abundance of solar energy that can be tapped. Indeed, the world's largest solar farm is currently found in California's Riverside County.

---

area, either in terms of the degree of dryness or its duration compared to average conditions. Hydrological droughts refer to lower than average water levels in river and surface water systems, as well as low groundwater levels, and often occur at the same time as meteorological drought. Agricultural drought relates to a lack of water at key points in the growing season, be it due to late onset or early cessation of rain, or due to meteorological drought. Farmers are often most concerned by agricultural droughts because they have an important socio–economic impact by reducing crop yields.

In the world's drylands, where the climate is inherently variable, drought is a common occurrence. The drylands are inhabited by more than two billion people, who suffer from some of the world's lowest levels of human well-being and highest incidences of poverty (Thomas, 2008). More people depend on the natural environment to meet their basic needs in drylands than in any other ecosystem (MA, 2005). And yet, the lack of rain and its unpredictability significantly limits dryland productivity and the extent to which people can depend upon their environment to provide for them in any given year. This means that people's livelihoods (the ways in which they make a living and give meaning to their lives), along with the ways in which they live, are strongly influenced by the climate. It also means that people who live in drylands have learned how to adapt their livelihoods over millennia to manage these types of conditions. This creates a rich source of locally held knowledge that could potentially be drawn upon to help manage future changes.

If we can learn from the collective experience of dryland dwellers, it may help people who face more frequent and severe droughts under climate change to prepare for the future. But we need to do this at the same time as feeding a rapidly growing global population on a land base that is shrinking due to conflicting land use interests (e.g. urbanization, conservation or mining) and ongoing sea level rises. We must also do this without compromising the medium- and long-term capacity of the land to provide the resources upon which populations in drylands and other biomes depend. Understanding the relationships between climate change and land degradation is therefore essential if the environment is to continue to provide humans with what we need for our survival and well-being long into the future.

## 1.6 Aims and structure of this book

This book provides new scientific insights and recommendations for policy and practice about how to anticipate, assess and adapt to land degradation and climate change. For the first time, we consider how vulnerable human populations and ecosystems around the world might be affected by the combined effects of land degradation and climate change, considering their exposure and sensitivity, and ways of adapting to interactions between these two processes.

We do this over the course of nine chapters:

- Policy context (Chapter 2): in this chapter we review each of the key international policy instruments relating to climate change and land degradation, and consider synergies between these and other policy instruments (for example relating to biodiversity). Chapter 2 concludes by considering how these instruments may be used to move towards a land degradation neutral situation where there is zero net land degradation.
- Conceptual and methodological frameworks (Chapter 3): based on a synthesis of conceptual frameworks this chapter proposes a methodological framework

for assessing the vulnerability of social-ecological systems to land degradation and climate change. Frameworks presented in this chapter are then used to structure the rest of the book.

- Exposure and sensitivity of provisioning ecosystem services (Chapter 4): this chapter focuses on assessing the vulnerability of provisioning ecosystem services to the effects of climate change and land degradation, with a particular focus on agricultural and livelihood systems.
- Exposure and sensitivity of other ecosystem services and feedbacks between climate change and land degradation (Chapter 5): this chapter considers the range of other types of ecosystem services that are likely to be affected by land degradation and climate change, including regulating, supporting and cultural services. The chapter outlines a range of feedbacks between different system components and climate and degradation processes.
- Responses (Chapter 6): this chapter considers how adaptive capacity can be enhanced to retain the integrity of ecosystems in regions affected by DLDD and maintain sustainable livelihoods in the face of the interactive effects of climate change and land degradation. It assesses how barriers to adaptation may be overcome to achieve a 'triple-win' scenario in the context of the three Rio Conventions, whereby adaptation addresses the impacts of climate change, land degradation and biodiversity loss.
- Monitoring and evaluating current and future effects of climate change and land degradation (Chapter 7): this chapter explores a range of monitoring and evaluation methods from both the environmental and social sciences. It considers methods for monitoring the current effects of land degradation and climate change, and reviews approaches for assessing likely future effects of climate change and land degradation.
- Monitoring and evaluating response options (Chapter 8): this chapter assesses approaches for evaluating response options, including methods for monitoring adaptation. It considers the role that political and institutional, economic and socio-technical factors are likely to play in creating constraints and trade-offs for adaptation. Methods are then presented for assessing the characteristics of adaptations that may make them more or less likely to be adopted, before considering methods for monitoring adaptation.
- Involving stakeholders (Chapter 9): this chapter considers evidence for multi-stakeholder approaches to anticipating, assessing and adapting to the challenges of climate change and land degradation. It considers how knowledge exchange and co-operation may be enhanced between local, scientific, NGO and policy communities, and critically assesses the role of locally held and scientific knowledge in developing responses to the combined effects of climate change and land degradation.
- Conclusion (Chapter 10): this chapter synthesises the key points and take home messages raised during the previous nine chapters, and sets out a forward-looking research agenda to address the key knowledge gaps in need of further research.

## Notes

1 Defined by MA (2005) as "a dynamic complex of plant, animal and micro-organism communities, and the non-living environment, interacting as a functional unit".
2 The biological, chemical, physical and hydrological processes through which ecosystems function, for example decomposition, dispersal and fluxes of nutrients and energy.
3 Defined as "the benefits provided by ecosystems that contribute to making human life both possible and worth living. The term 'services' is usually used to encompass the tangible and intangible benefits that humans obtain from ecosystems, which are sometimes separated into 'goods' and 'services'" (UK National Ecosystem Assessment, 2011).
4 In this book, the terms DLDD and "land degradation" are used interchangeably, implicitly including desertification in dryland contexts. These terms are defined in depth in Section 1.1.
5 It should be noted that the MA includes hyper-arid lands in its definition of drylands.

## References

Ballayan, D. 2000. Soil degradation. FAO Chapter 6: Biodiversity and Land degradation. *ESCAP Environment Statistics Course*, 24 pp.

Burke, E.J., Brown, S.J., Christidis, N. 2006. Modeling the recent evolution of global drought and projections for the Twenty-First century with the Hadley Centre Climate Model. *Journal of Hydrometeorology* 7: 1113–1125.

Cantagallo, J.E., Chimenti, C.A., Hall, A.J. 1997. Number of seeds per unit area in sunflower correlates well with the photothermal quotient. *Crop Science* 37: 1780–1786.

Chun, J.A., Wang, Q.G., Timlin, D., Fleisher, D., Reddy, V.R. 2013. Effect of elevated carbon dioxide and water stress on gas exchange and water use efficiency in corn. *Agricultural and Forest Meteorology* 151: 378–384.

Dasgupta, S., Kamal, F.A., Akhter, K., Zahirul, H., Choudhury, S., Nishat, A. 2014a. River salinity and climate change: evidence from coastal Bangladesh. Policy Research working paper; no. WPS 6817. World Bank Group: Washington, D.C., USA. http://documents.worldbank.org/curated/en/2014/03/19299368/river-salinity-climate-change-evidence-coastal-bangladesh

Dasgupta, S., Hossain, Md. M., Mainul, H., Wheeler, D. 2014b. Climate change, soil salinity, and the economics of high-yield rice production in coastal Bangladesh. Policy Research working paper; no. WPS 7140. Paper is funded by the Knowledge for Change Program (KCP). World Bank Group: Washington, D.C., USA. http://documents.worldbank.org/curated/en/2014/12/23056341/climate-change-soil-salinity-economics-high-yield-rice-production-coastal-bangladesh

Dasgupta, S., Hossain, Md. M., Mainul, H., Wheeler, D. 2014c. Climate change, groundwater salinization and road maintenance costs in coastal Bangladesh. Policy Research working paper; no. WPS 7147. Paper is funded by the Knowledge for Change Program (KCP). World Bank Group: Washington, D.C., USA. http://documents.worldbank.org/curated/en/2014/12/23129701/climate-change-groundwater-salinization-road-maintenance-costsbrin-coastal-bangladesh

D'Odorico, P., Bhattachan, A., Davis, K.F., Ravi, S., Runyan, C.W. 2013. Global desertification: drivers and feedbacks. *Advances in Water Resources* 51: 326–344.

Dregne, H.E. 1983. *Desertification of Arid Lands*. Harwood Academic Publishers: Switzerland.

Dregne, H.E., Chou, N.T. 1992. Global desertification dimensions and costs. In: *Degradation and Restoration of Arid Lands*, Dregne H.E. (ed.). Lubbock, TX: Texas Tech. University, pp. 249–282.

EHRS (Ethiopian Highlands Reclamation Study). 1984. Annual research report 1983–1984. EHRS: Addis Ababa, Ethiopia.

Favretto, N., Stringer, L.C., Dougill, A.J. 2014. Unpacking livelihood challenges and opportunities in energy crop cultivation: perspectives on *Jatropha curcas* projects in Mali. *Geographical Journal* 180: 365–376.

Holling, C.S. 1986. The resilience of terrestrial ecosystems, local surprise and global change. In: *Sustainable Development of the Biosphere*, Clark, W.C., Munn, R.E. (eds). Cambridge University Press: Cambridge, pp. 292–317.

IPCC (Intergovernmental Panel on Climate Change). 2001. *Climate Change 2001: Impacts, adaptation, and vulnerability*. Cambridge University Press: Cambridge.

IPCC. 2007. Climate Change 2007: The physical science basis. In: *Contribution of Working Group I to the Fourth Assessment Report of the Intergovernmental Panel on Climate Change*, Solomon, S., Qin, H.D., Manning, M., Chen, Z., Marquis, M., Averyt, K.B., Tignor, M., Miller, H.L. (eds). Cambridge University Press: Cambridge, United Kingdom and New York, NY, USA, 996 pp.

Kaminski, K.P., Korup, K., Nielsen, K.L., Liu, F.L., Topbjerg, H.B., Kirk, H.G., Andersen, M.N. 2014. Gas-exchange, water use efficiency and yield responses of elite potato (*Solanum tuberosum* L.) cultivars to changes in atmospheric carbon dioxide concentration, temperature and relative humidity. *Agricultural and Forest Meteorology* 187: 36–45.

Kasperson, J., Kasperson, R., Turner, B. 1995. Regions at risk. United Nations University Press. Available online at: www.unu.edu/unupress/unupbooks/uu14re/uu14re00.htm

Keenan, T.F., Hollinger, D.Y., Bohrer, G., Dragoni, D., Munger, J.W., Schmid, H.P., Richardson, A.D. 2013. Increase in forest water-use efficiency as atmospheric carbon dioxide concentrations rise. *Nature* 499: 324–327.

Le Houérou, H.N. 1996. Climate change, drought and desertification. *Journal of Arid Environments* 34: 133–185.

Lepers, E., Lambin, E.F., Janetos, A.C., DeFries, R., Achard, F., Ramankutty, N., Scholes, R.J. 2005. A synthesis of information on rapid land-cover change for the period 1981–2000. *BioScience* 55: 115–124.

Mabbutt, J.A. 1984. A new global assessment of the status and trends of desertification. *Environmental Conservation* 11: 103–113.

MA (Millennium Ecosystem Assessment). 2005. *Ecosystems and Human Well-being: Current State and Trends Assessment*. Island Press: Washington, D.C., USA.

Morgan, R.P.C. 2005. *Soil Erosion and Conservation*, 3rd edition. National Soil Resources Institute, Cranfield University. Blackwell Publishing Ltd.

Reed, M.S., Stringer, L.C., Dougill, A.J., Perkins, J.S., Atlhopheng, J.R., Mulale, K., Favretto, N. 2015. Reorienting land degradation towards sustainable land management: linking sustainable livelihoods with ecosystem services in rangeland systems. *Journal of Environmental Management* 151: 472–485.

Scalenghe, R., Marsan, F.A. 2009. The anthropogenic sealing of soils in urban areas. *Landscape and Urban Planning* 90: 1–10.

Seager, R., Ting, M.F., Held, I., Kushnir, Y., Lu, J., Vecchi, G., *et al.* 2007. Model projections of an imminent transition to a more arid climate in southwestern North America. *Science* 316: 1181–1184.

Selassie, Y.G., Amede, T. 2014. Investing in land and water management practices in the Ethiopian Highlands: short- or long-term benefits? In: *Challenges and Opportunities for Agricultural Intensification of the Humid Highland Systems of Sub-Saharan Africa*, Vanlauwe, B. *et al* (eds). Springer International Publishing Switzerland 2014, pp.105–114.

Thomas, R.J. 2008. 10th Anniversary review: addressing land degradation and climate change in dryland agroecosystems through sustainable land management. *Journal of Environmental Monitoring* 10: 595–603.

Travasso, M.I., Magrin, G.O., Rodriguez, G.R., Boullon, D.R. 1999. Climate Change Assessment in Argentina: II. Adaptation strategies for agriculture. In: *Food and Forestry: Global Change and Global Challenge*. GCTE Focus 3 Conference. Reading: UK.

Turner, B.L., II, Benjamin, P. 1993. Fragile lands and their management. In: *Agriculture, Environment and Health: Towards Sustainable Development into the 21st century*, Ruttan V.W. (ed.). University of Minnesota Press: Minneapolis.

UK National Ecosystem Assessment. 2011. The UK National Ecosystem Assessment: Technical report. UNEP-WCMC: Cambridge.

UNCCD and CFC. 2009. *African Dryland Commodity Atlas*. United Nations Convention to Combat Desertification and the Common Fund for Commodities, Bonn, Germany. Available online at: www.unccd.int/en/resources/publication/Pages/Other-publications. aspx

UNCCD. 1994. *United Nations Convention to Combat Desertification in Those Countries Experiencing Serious Drought and/or Desertification Particularly in Africa: Text with Annexes*. UNEP: Nairobi.

UNFCCC (United Nations Framework Convention on Climate Change). 1992. *United Nations Framework Convention on Climate Change: Text with Annexes*. Available online at: http://unfccc.int/key_documents/the_convention/items/2853.php

Warren, A.S. 2002. Land degradation is contextual. *Land Degradation and Development* 13: 449–459.

Worster, D. 2004. (1979; 25th anniversary edition) *Dust Bowl: The southern plains in the 1930s*. Oxford University Press.

# 2

# POLICY CONTEXT

This chapter sets out the international policy context for climate change, land degradation and desertification, focusing on the key international instruments designed to tackle the issues: the UN Framework Convention on Climate Change (UNFCCC) and the UN Convention to Combat Desertification (UNCCD). These agreements, together with the Convention on Biological Diversity (CBD), are collectively known as the three "Rio Conventions". They help to support the implementation of Agenda 21, the main outcome document from the UN Conference on Environment and Development (UNCED), held in Rio de Janeiro, Brazil, in 1992 (Box 2.1). This chapter also provides a brief outline of the processes that led to the current arrangements of roles and responsibilities of policy actors at the various scales of implementation, and sets out the policy need to examine the links between climate change and land degradation.

## 2.1 The UNFCCC

Global recognition that the climate was changing far pre-dated the negotiation of the UNFCCC. The severe droughts and famines of the 1960s and 1970s, which bore considerable human costs, triggered a range of international events, convened to kick-start discussions about how the human population could better manage its relationship with the environment. One such event was the UN Conference on Desertification (UNCOD) held in 1977. Another was the First World Climate Conference (FWCC), arranged in 1979 by World Meteorological Organization (WMO), United Nations Environment Programme (UNEP), Food and Agriculture Organization (FAO), United Nations Education, Scientific and Cultural Organization (UNESCO) and the World Health Organization (WHO). The aim of the FWCC was to identify the state of knowledge about the climate and to examine the impacts of climate variability and change on human society. The meeting ultimately led to the creation of the Intergovernmental Panel on Climate Change (IPCC) in 1988.

## BOX 2.1: THE RIO CONVENTIONS

The three Rio Conventions – on Biodiversity, Climate Change and Desertification – were all discussed during the 1992 United Nations Conference on Environment and Development (also known as the Earth Summit), held in Rio de Janeiro, Brazil. Agenda 21 is the main sustainable development framework document that was agreed upon at the Earth Summit. Each Rio Convention offers a way to support the sustainable development goals of Agenda 21, while also addressing issues that are interdependent within the same ecosystems.

The CBD focuses on "conserving biological diversity, the sustainable use of its components, and the fair and equitable sharing of the benefits arising from commercial and other utilization of genetic resources". It applies to all ecosystems, species and genetic resources.

The UNCCD focuses on combatting desertification and mitigating the effects of drought

> in countries experiencing serious drought and/or desertification, particularly in Africa, through effective actions at all levels, supported by international co-operation and partnership arrangements, in the framework of an integrated approach which is consistent with Agenda 21, with a view to contributing to the achievement of sustainable development in affected areas.

The UNFCCC sets out a framework for governments to address the challenges presented by climate change. Its objectives are to

> stabilize greenhouse-gas concentrations in the atmosphere at a level that would prevent dangerous anthropogenic interference with the climate system, within a time-frame sufficient to allow ecosystems to adapt naturally to climate change; to ensure that food production is not threatened; and to enable economic development to proceed in a sustainable manner.

The inherent links between the three Rio Conventions means that there is some danger of duplication of activities but also considerable potential for synergy. In August 2001, a Joint Liaison Group (JLG) between the Rio Conventions was established for exchanging information, exploring opportunities for synergistic activities and increasing coordination. Although some authors have stated that the JLG could do much more to harness synergy, particularly when it comes to monitoring and data sharing (Bauer and Stringer, 2009), their collaborative efforts are starting to see more joined-up working and the development of a number of shared implementation activities.

Modified from: www.cbd.int/rio/

The IPCC was charged with preparing, on the basis of available scientific information, assessments on all aspects of climate change and its impacts, in order that realiztic responses may be formulated. The IPCC's current purpose, as defined in the Principles Governing IPCC Work, is:

> ...to assess on a comprehensive, objective, open and transparent basis the scientific, technical and socio-economic information relevant to understanding the scientific basis of risk of human-induced climate change, its potential impacts and options for adaptation and mitigation. IPCC reports should be neutral with respect to policy, although they may need to deal objectively with scientific, technical and socio-economic factors relevant to the application of particular policies.
>
> *(IPCC, 2016)*

The scientific findings of the first IPCC Assessment Report presented in 1990 drew attention to the global nature of the climate change challenge and was a key stimulus for the creation of the UNFCCC. The UNFCCC was opened for signature at the UN Conference on Environment and Development (the Earth Summit) held in Rio de Janeiro, Brazil, in 1992. It entered into force on 21 March 1994 and today has 195 Parties. The objective of the UNFCCC is to stabilize greenhouse gas concentrations "at a level that would prevent dangerous anthropogenic (human induced) interference with the climate system" (UNFCCC, 1992). It notes that "such a level should be achieved within a time-frame sufficient to allow ecosystems to adapt naturally to climate change, to ensure that food production is not threatened, and to enable economic development to proceed in a sustainable manner" (UNFCCC, 1992). The agreement acknowledges that industrialized countries belonging to the Organization for Economic Cooperation and Development (OECD) are the source of the majority of past and present emissions. As such, the treaty placed responsibility on those countries (known as Annex I countries) to take more substantial steps to reduce their emissions to 1990 levels by the year 2000.

By 1995 however, it was realized that the provisions in the Convention to reduce emissions were insufficiently strong to achieve what was considered necessary to stabilize global greenhouse gas emissions. In response, negotiations started to develop a legally binding protocol that requires developed countries to meet particular emission reduction targets and plan for adaptation to future climate change (Box 2.2). The Kyoto Protocol was adopted in 1997 and its first commitment period ran from 2008 to 2012. The second commitment period began on 1 January 2013 and runs until 2020. Of the 195 Parties to the Convention, 192 Parties have also signed up to the Kyoto Protocol.

## 2.2 The UNCCD

The United Nations Convention to Combat Desertification (UNCCD) of 1994 came into force in 1996 following its fiftieth ratification. The route to the UNCCD's

## BOX 2.2: MITIGATION AND ADAPTATION WITHIN THE UNFCCC

With the goal of the UNFCCC focused on greenhouse gas emission reductions, the majority of the world's attention was initially directed towards mitigation activities. The risks associated with inaction were obvious: emissions had to be lowered or absorbed by managing carbon sinks in order to reduce the magnitude and rate of climate change, otherwise humanity faced a dangerous climatic outcome. Those who called for a similar emphasis on adaptation (and the necessary funds to support it) were viewed as being defeatist and pessimistic. At the same time, OECD countries were keen to avoid having to debate questions of liability, compensation, equity and fairness (Paavola *et al.*, 2006). It was only with the 2001 Marrakesh Accords to the Bonn Agreements under the UNFCCC that the situation became more balanced, as policymakers attending UNFCCC COP7 finally recognized that mitigation and adaptation can be synergetic and that they both play a key part in the management of future climate change.

Since 2001, aside from Annexe I countries being required to report on their policies and measures (including issues governed under the Kyoto Protocol), they are also obliged to submit annual inventories of all their emissions from 1990 to date. Developing countries are required to report more generally on the steps they are taking to both address climate change and to adapt to its impacts. However, reporting frequencies are highly dependent on countries' abilities to harness funds and resources to enable them to report. For the world's Least Developed Countries (LDCs), which it is argued have contributed least in the way of emissions, the 2001 agreements paved the way for a special focus on climate impacts in LDCs, which includes the development of National Adaptation Plans of Action (NAPAs) (Stringer *et al.*, 2009). NAPAs allow LDCs to identify the most urgent activities necessary to address their most immediate needs regarding adaptation to climate change, which, in the absence of action, could increase their vulnerability or lead to increased future costs. The UNFCCC website notes that in the majority of cases, the most urgent issues relate to agriculture, food security, water resources, coastal zones, and early warning and disaster risk reduction, so strong prominence is given to the natural resource base sectors. The process taken in the development of NAPAs encompasses the need for local-level inputs, and recognizes the value of locally held knowledge in informing adaptations. Once a LDC has prepared its NAPA and submitted it to the UNFCCC secretariat, it is eligible to apply for resources from the Least Developed Countries Fund (LDCF) in order to move towards implementation.

presence on the international political stage was not, however, smooth. The idea for a global treaty on land degradation and desertification initially came from the African states in the run up to the 1992 Earth Summit. Aside from the Sahel droughts and famine of the 1970s, across Africa, countries were concerned that the world's growing focus on climate change and biodiversity would take attention away from the challenges that they considered were a key obstacle to the continent's sustainable development. However, not all countries were convinced of the need for a global treaty on desertification. Many developed countries, especially the European Union's member states of the time, considered that desertification was not a global problem, as it was experienced most acutely at the local level. They initially opposed an international agreement, arguing that it was a local problem resulting from poor land use decision-making by land users. They also referred back to the policy efforts of the 1970s to address dryland problems in particular, and recalled the general lack of policy effectiveness and success at that time (Box 2.3).

It was only during the final stages of the Rio summit that the developed countries were willing to concede. Scientific advances that had begun to unravel the complexities of desertification gave the African argument a leverage point that they used to persuade the developed world to back down. African governments argued that local land management decisions everywhere were affected by wider drivers, such as global trade, climate, human migration and technology and that this made the problem a global issue. Land degradation and desertification occurring at the local level in Africa would affect the provision of dryland commodities on world markets. At the same time, the first edition of UNEP's World Atlas of Desertification had recently been published, and drew attention to the magnitude of land areas and people who were experiencing or at risk of experiencing desertification and soil degradation. Eventually, the position of the African countries prevailed, and the UN General Assembly established the Intergovernmental Negotiating Committee on the Desertification Convention.

The resulting Convention text presents a multi-scale, participatory approach, empowering all stakeholders to work together to tackle DLDD. The mission of the UNCCD is:

> To provide a global framework to support the development and implementation of national and regional policies, programmes and measures to prevent, control and reverse desertification/land degradation and mitigate the effects of drought through scientific and technological excellence, raising public awareness, standard setting, advocacy and resource mobilization, thereby contributing to poverty reduction.
>
> *(UNCCD, 2008)*

The Convention recognizes land users as key actors in addressing DLDD and appreciates the value of locally held knowledge in developing solutions. At the same time, the UNCCD requires governments to work independently to ensure the integration of DLDD issues into national policy- and decision-making, as well as working together to share experiences, knowledge, technologies and other resources.

## BOX 2.3: POLITICAL EFFORTS TO COMBAT DESERTIFICATION IN THE 1970S–80S

The first efforts to bring desertification to the global stage followed the severe drought and famine in Africa's Sudano-Sahel region between 1968 and 1974. The UNEP responded to this situation by convening the United Nations Conference on Desertification (UNCOD), held in 1977. The objective of the UNCOD was to advance scientific knowledge about desertification, drought and their socio-economic impacts on dryland populations, while at the same time stimulating development and mitigation of desertification in degraded areas. Following the UNCOD, the UN General Assembly adopted a UN Plan of Action to Combat Desertification (PACD). This provided UNEP with the mandate to coordinate activities with a view to controlling desertification by the millennium. The PACD proposed that:

1) National government institutions should be created in order to combat desertification;
2) Regional Economic Commissions should arrange and coordinate the implementation of transboundary projects to combat desertification, convene other regional fora (conferences, seminars, workshops), undertake inter-regional studies, and also set up regional training centres to tackle the problem;
3) UN Agencies and Organizations should participate in a plan of action based on requests for assistance received from governments, while UNEP would collaborate with other agencies to identify particular actions, mobilize resources and coordinate projects to combat desertification (FAO, 1993).

Although involving actors at multiple scales, the PACD's approach was initiated in a top-down way, yet at the same time placing the onus on governments at the national level to request assistance. By the run up to the Earth Summit, a UNEP evaluation had highlighted that progress with the PACD had been rather minimal, and there was a general realization that tackling desertification was more complex than it initially had seemed. Up until that point, African governments and international donors had not made desertification a priority. Financial support from governments had not been forthcoming; national development plans had not taken anti-desertification activities into account; and the policy and legislative environment had not created an enabling context to tackle the human drivers of the desertification challenge. Developed countries initially highlighted these challenges as a way to oppose the negotiation of the UNCCD.

The history of the negotiation of the Convention is partly reflected in its full title "The United Nations Convention to Combat Desertification and Drought in Those Countries Experiencing Serious Drought and/or Desertification, Particularly in Africa". This title recognizes that the specific challenges of DLDD are critical for the sustainable development of the African continent, while the Regional Annex for Africa and Annexes for other regions of the world experiencing shared degradation problems, provide a route to identifying the core issues of focus for cooperation activities.

In 2016, twenty-two years after the UNCCD was first ratified, it is widely supported as an instrument for tackling DLDD, whilst contributing towards sustainable development and poverty alleviation. It has almost universal membership (195 Parties), with the most recent country to join being South Sudan. Box 2.4 describes the responsibilities of the Parties to the UNCCD in more detail.

## 2.3 Implementing the UNCCD through its 10-year strategy

The Parties to the UNCCD adopted a 10-year strategy for 2008–2018 at the eighth Conference of the Parties (COP8) in Madrid in 2007, to help address the Convention's key challenges (ICCD/COP(8)/16/Add.1). The strategy contains four objectives (with associated expected impacts): (i) to improve the living conditions of affected populations; (ii) to improve the conditions of affected ecosystems; (iii) to contribute towards global conservation and sustainable use of biodiversity and climate change mitigation; and (iv) to build partnerships between national and international actors to implement the Convention. It also has five operational objectives (with associated expected impacts): (i) advocacy, awareness raising and education; (ii) to create an enabling policy environment for promoting solutions to land degradation; (iii) to become a global authority on scientific knowledge about land degradation; (iv) to build capacity for reversing land degradation; and (v) to target and co-ordinate financial and technical resources for the Convention. Parties to the UNCCD are expected to align their NAPs and other implementation activities with the 10-year strategy, ensuring that they address the five operational objectives. An ambitious target of 80 per cent NAP alignment was set for 2014. However, by December 2015, an estimated 64 per cent of Parties will have achieved alignment.

At COP8, a decision was taken to strengthen the scientific basis of the work of the Convention as it pursues its 10-year strategy. This move was welcomed by the scientific community, as well as NGOs and other UNCCD stakeholders. The ways in which science had been channeled into UNCCD decision making via its Committee on Science and Technology (CST) had long been criticized, predominantly for three reasons: (i) scientific experts providing advice to the UNCCD's CST were usually government-nominated rather than independent; (ii) emphasis had been placed on regional representation and balance rather than on the competence and credibility of scientific experts; and (iii) the breadth of disciplinary input required to tackle the complex problems of DLDD had not always been sufficiently recognized, with socio-economic expertise often being neglected. While the IPCC

## BOX 2.4: RESPONSIBILITIES OF PARTIES TO THE UNCCD

Although the UNCCD is legally binding, there are no automatic sanctions or punishments should Parties fail to meet their obligations, as the agreement is not enforced. This is the case for many Multi-lateral Environmental Agreements (MEAs). Instead, Parties meet their obligations in order not to 'lose face' within the international community. Also, by complying with MEAs, countries are often able to access otherwise restricted sources of funding. There are four main types of obligations under the UNCCD and its regional implementation annexes. First, there are common obligations of all Parties, whether or not they consider themselves affected by desertification. These obligations link mostly to international cooperation in the implementation of the UNCCD and focus attention towards the

> collection, analysis and exchange of information, research, technology transfer, capacity building and awareness building, the promotion of an integrated approach in developing national strategies to combat desertification, and assistance in ensuring that adequate financial resources are available for programmes to combat desertification/land degradation and mitigate the effects of drought.
>
> *(UNCCD, 2015)*

All Parties are required to report regularly on the measures they have taken to implement the UNCCD, in line with its Article 26.

Second, those Parties that consider themselves to be affected by DLDD and which are covered by one of the five Regional Annexes (I: Africa, II: Asia, III: Latin America and the Caribbean, IV: Northern Mediterranean and V: Central and Eastern Europe) are obliged to develop National Action Programmes (NAPs) to combat desertification. NAPs should set out the key DLDD issues within each country, as well as identify the major strategies, programmes and projects to address those challenges. Those Parties that prepare a NAP are obliged to report regularly on the progress made with implementation. The Regional Annexes are also intended to foster cooperation at regional and subregional levels, as Parties within a particular region often suffer the same or similar DLDD challenges.

Third, Parties that do not align with a Regional Annex but which consider themselves affected by DLDD have the possibility to prepare an action programme following the provision of the UNCCD, or to establish strategies and priorities to address the particular DLDD issues they face.

Fourth, developed-country Parties have particular obligations to provide support to affected countries, particularly in the developing world. Such support can take the form of monetary resources, technology, knowledge and know-how.

had been established before the UNFCCC, it was nevertheless perceived to provide the UNFCCC's Subsidiary Body for Scientific and Technological Advice with the scientific evidence base deemed necessary to make informed policy decisions. Up to the point of COP8, the UNCCD had lacked such a source of information, with the CST having to rely on a roster of (government-nominated) experts for *ad hoc* input.

Part of the strengthening of scientific input required by the COP8 decision was the introduction of Scientific Conferences. To date, the UNCCD's CST has held three Scientific Conferences, with the most recent taking place in 2015. A key conclusion from the first conference was the need to combine biophysical assessments of land degradation with an appreciation of stakeholder perceptions of changes in the capacity for the land to support their livelihoods (Winslow *et al.*, 2011). To do this, the conference recommended the use of integrated assessment modelling employing a flexible range of indicators that detect information at different scales, which can draw on both locally held and scientific knowledge about land degradation processes, severity and extent. In this way, it was argued that monitoring and assessment could feed into decision-making at national and sub-national scales, which could enhance the capacity for ecosystems and populations in regions affected by DLDD to adapt to and mitigate land degradation. These themes will be revisited in relation to both land degradation and climate change in Chapter 3.

The second conference analyzed the economic and social costs of land degradation as well as the benefits of sustainable land management. The conference identified a range of policy mechanisms that could incentivize more sustainable management, in an attempt to reach land degradation neutrality (defined as a zero net degradation state where the rate of land degradation is equal to the rate of land restoration). The third conference focused on "Combating desertification, land degradation and drought for poverty reduction and sustainable development – the contribution of science, technology, traditional knowledge and practices" and took place at the fourth special session of the Committee on Science and Technology (CST S-4) from 9 to 12 March 2015 in Cancun, Mexico. To inform this conference, an Impulse Report was commissioned on links between land degradation and climate change, which was the starting point for this book.

## 2.4 Harnessing synergy in tackling climate change and land degradation through the Rio Conventions

As the UNCCD's conference themes have developed over time in response to evolving policy needs, the importance of harnessing synergy between the Rio Conventions has become more widely appreciated. The shift away from viewing climate change and biodiversity loss as purely biophysical phenomena and the acknowledgement of land degradation as a problem with global drivers and consequences brings international policy more in line with local realities that implicitly take this kind of complexity into account. For example, in addition to responding to the challenges of land degradation and climate change, ecosystems and people in regions affected by DLDD must deal with other stresses such as changing market prices and trade conditions, changing policies, population and demographic changes, as well as problems of disease (affecting plants, animals and humans). Land degradation directly contributes to ongoing losses of biodiversity, while land degradation processes and impacts can be exacerbated by climate change in a range of complex and often unpredictable ways (MA, 2005; Thomas, 2008).

By placing climate change, land degradation and biodiversity in the broader sustainable development context, it is also possible to identify links with other global challenges, as well as opportunities for synergy in tackling the issues. Such broader development challenges include, for example, achieving food security (Box 2.5); the need to understand and manage the relationships between resource scarcity, migration and conflict (often related to land and water insecurity, inequality and a lack of livelihood opportunities); problems of biodiversity loss (species other than humans need to adapt as well, while genetic diversity – the amount of genetic variation within a species – needs also to be maintained); the global trend towards changing land use, as policy, subsidies and private sector investment increase support for the widespread plantation of non-food biomass such as energy crops, tobacco, rubber and cotton, many of which are cultivated in dry areas with the use of irrigation; and the development of a greener economy.

These links present numerous opportunities for synergy in the implementation of all three Rio Conventions, and in addressing the root causes of the problems. Harnessing synergy reduces the potential for both conflicting activities and duplication of efforts between the independent initiatives of each Convention. Approaches that seek out synergy can also enable resources to be used more efficiently.

A number of mechanisms have been employed to advance cooperation between the Conventions. Collaborative approaches have been developed in several crosscutting areas, including: knowledge management, research and monitoring (e.g. the JLG and the work of the Intergovernmental Platform on Biodiversity and Ecosystem Services, which is expected to support the implementation of strategic plans of the three Rio Conventions), joint work programmes (e.g. the CBD's (2011) Strategic Plan for Biodiversity 2011–2020) and joint financing initiatives (e.g. the UNFCCC's Green Climate Fund which can be used to fund projects that reduce greenhouse gas emissions whilst reversing land degradation). All three Rio Conventions have also actively participated in initiatives such as the International Year of Deserts and Desertification (2006) (see Box 2.6), the International Year of Soil (2015) and the International Decade on Deserts and Desertification (2010–2020).

In each of the three Rio Conventions, significant consideration has been given to the role of science and other forms of knowledge to inform policy. This led first to the engagement of the IPCC to provide advice to the UNFCCC, then to the development of the Intergovernmental Platform on Biodiversity and Ecosystem Services (IPBES) linked to the CBD, and most recently to the establishment of a Science Policy Interface (SPI) under the UNCCD. The SPI is a scientific advisory mechanism that includes both governmental and independent experts of equal number, thus avoiding many of the criticisms levelled at earlier mechanisms for the provision of scientific advice to the UNCCD. It also comprises observers from a civil society

## BOX 2.5: FOOD SECURITY AS A CROSS-CUTTING CHALLENGE FOR THE RIO CONVENTIONS

Whether from climate change, land degradation or a combination of the two, it would appear that without substantial adaptations, climate change is likely to increase food insecurity in regions affected by DLDD across the developing world that are experiencing rapid population growth. Indeed, Parry *et al.* (1999, 2005) have suggested that Africa is at greatest risk from the effects of climate change on food production and hunger, and predicted that there are likely to be millions more people at risk of hunger there by the 2080s. On the other hand, Fischer *et al.* (2008) suggest that reductions in yield in the developing world may be offset by yield increases of similar magnitude in the developed world (5 to 10% by 2050) as climate change makes conditions more favourable for agricultural production there. However, it is unlikely that these increases in yield for developed nations will provide any more than a small part of the solution for increasingly food insecure developing world countries whose population growth is likely to far outstrip these productivity gains. Furthermore, land degradation is likely to significantly exacerbate food insecurity as it interacts with climate change (Gregory *et al.*, 2005). These issues are considered in greater detail in Chapter 4.

In 1996, the World Food Summit defined food security as "when all people at all times have access to sufficient, safe, nutritious food to maintain a healthy and active life". This definition first encompasses issues of availability. A food secure situation needs sufficient quantities of food to be consistently available. This requires reliable production, which depends upon a suitable climate, sufficient good quality land and the biodiversity necessary for processes such as pollination. Second, the World Food Summit definition considers questions of access: for people to be food secure requires not just the presence of food but that everyone has enough resources (both physical and economic) to obtain appropriate, nutritious foods. Finally, the definition encompasses the use of food, as well as adequate water and sanitation – this is vital in terms of people living healthy and active lives.

Climate change and land degradation affect each aspect of the definition and need to be viewed in the context of several key challenges, including distribution, production and sovereignty. Some commentators argue that there is already enough food in the world to feed everyone at present and that the main problem is one of distribution, with the challenge being largely about getting it to the areas of the world that are most in need. In the future, however, projected global populations and their associated food security needs will not be met by current levels of production. This is particularly critical when we consider that people need to be fed from a shrinking land base due to degradation and climate-change-related sea level rise.

Some estimates suggest that food production will need to increase by 70 per cent of current levels over the next 40 years and that a new "Green Revolution" is required. How the climate changes and the extent to which land is degraded perhaps most clearly plays into this concern, as it will affect both the amount of food produced and the ways in which it needs to be processed, stored and distributed.

The original Green Revolution improved the yields of wheat, rice and other staple crops mostly in countries such as India and Mexico. It involved huge investments in plant breeding research and development during the 1960s to the 1980s, coupled with the diffusion of innovations such as agricultural technology and hybrid seeds and fertilizers. Arguably, the Green Revolution may have produced more food, but because it only addressed part of the problem (production) and failed to address issues of access, there are still many people without sufficient nutritious food today. Studies also show that the dominance of Western technologies in the Green Revolution not only led to biodiversity loss and soil erosion, but also ignored the agricultural adaptations and local knowledge of generations of farmers with considerable know-how and experience, and neglected to consider the socio-cultural context of technological application.

Fears of another "partial" Green Revolution are in line with the growing disillusionment with the current global food system. Increasingly, there is realization that the dominant food security discourse places undue emphasis on commercial efficiency and production, and that it is outcome- rather than process-orientated, with important implications for long-term sustainability. It is felt to support large-scale, agro-industrial production based on specialized crops, land concentration and the liberalization of trade, and in doing so, draws attention away from associated issues such as the dispossession of small-scale producers and the negative environmental impacts of agro-industrial production. In response, food sovereignty has started to rise up the political agenda. This concept extends beyond food security to consider that those groups who produce, distribute and consume food should have a greater say in the mechanisms and policies that shape food security. In this way, it relates to questions of where food has come from and how it is produced. Food sovereignty is therefore an important consideration for initiatives that aim to tackle both land degradation and climate change and an important concern for both the UNCCD and UNFCCC.

organization, an international organization and a UN organization[1]. The SPI is responsible for presenting scientific inputs (including those from the UNCCD Scientific Conferences) to the CST, in order for the CST to make recommendations to the COP. The SPI also liaises with the scientific bodies of the other Rio Conventions.

The SPI and CST provided inputs to the IPBES at both the scoping stage of a Thematic Assessment on Land Degradation and Restoration to meet objective 3(b)(i) of the IPBES work programme[2], and through the nomination of independent scientific experts as authors of the assessment. The Land Degradation and Restoration Assessment is intended to provide a knowledge base for future policies addressing land degradation and restoration of degraded land for the CBD, UNCCD and any other bodies for whom the information is useful. The CST proposed that the assessment covers:

- the global status of and trends in land degradation, by region and land cover type;
- the effect of degradation on biodiversity values, ecosystem services and human well-being; and
- the state of knowledge, by region and land cover type, of ecosystem restoration extent and options.

Processes like the IPBES assessment, alongside the outcomes from the UNCCD Scientific Conferences are providing evidence on the processes through which land degradation, climate change and biodiversity are linked. Over time, these processes will help to strengthen the relationship between each of the Rio Conventions, contributing the aims of each treaty, as well as being informed by the work of other bodies working in related areas, such as the Intergovernmental Technical Panel on Soils (ITPS).

Synergy is also possible through joint efforts to implement the Sustainable Development Goals (SDGs). Goal 15.3 seeks to "combat desertification, and restore degraded land and soil, including land affected by desertification, drought and floods, and strive to achieve a land degradation neutral world", while Goal 13 is to "Take urgent action to combat climate change and its impacts". Indeed, the idea of land degradation neutrality gained traction as part of "The Future We Want" outcome document (UNGA, 2012) adopted at the United Nations Conference on Sustainable Development (Rio+20), and builds on existing environmental goals, such as Agenda 21 and the MDG. It is a framework for action, which seeks, via the UNCCD's strategic plan, to reduce degradation and scale up restoration activities from community to landscape scales.

## 2.5 Land degradation neutrality

The concept of Land Degradation Neutrality (LDN) accepts that further land degradation is inevitable, in light of the multiple pressures placed upon the natural resource base from the growing population and in the context of climate change (Tal, 2015). However, it places more of an emphasis on the restoration of degraded areas (and notes the benefits of doing so) than has been the case in the past (Barkemeyer et al., 2015). It also further supports the UNCCD's case for the need for global action to tackle land degradation in its various forms across the world and underscores the importance of the phrase "think global, act local", which became widely used after the 1992 Earth Summit (Chasek et al., 2015).

## BOX 2.6: INTERNATIONAL YEAR OF DESERTS AND DESERTIFICATION

The United Nations General Assembly's 58th Ordinary Session in 2003 declared 2006 the International Year of Deserts and Desertification (IYDD). This decision was adopted as a result of concerns about "the apparent exacerbation of desertification, particularly in Africa, and the negative implications it has for achieving the Millennium Development Goals (MDGs)" (UNGA, 2004). The General Assembly's support for the IYDD was provided with a view to enhancing progress towards MDG 7: to ensure environmental sustainability. IYDD was considered a way of "raising public awareness about desertification" and "helping to protect dryland biodiversity and the knowledge and traditions of the people whose everyday lives are affected by desertification" (UNGA, 2004). It provided a platform to celebrate the landscapes, cultures, traditions and knowledges of dryland inhabitants, while also raising awareness and the public visibility of desertification.

The UNCCD Secretariat used the abbreviation LAND to structure four objectives for the IYDD:

- long-term oriented implementation of the UNCCD;
- awareness of the implications of desertification;
- networking with all stakeholders; and
- dissemination of information relating to the UNCCD.

The events and activities that ensued took the form of meetings (including scientific and political conferences, workshops, lectures and seminars), cultural events (photographic and art exhibitions, dance shows, films, musical events and the issuing of commemorative coins and stamps), as well as public awareness-raising activities linked to new outreach and educational material, campaigns and competitions.

The UNCCD received wide international support from other UN bodies and Rio Conventions. For example, the CBD announced dryland biodiversity as the theme for International Biodiversity Day on 22 May 2006, while World Environment day on 5 June 2006 was marked by numerous global events, accompanied with the slogan "Don't Desert Drylands". The combination of different activities viewed the IYDD as an opportunity to catalyse the generation and diffusion of knowledge about drylands, and develop joint goals for the future, as well as increase participation in the fight against global desertification and contribute towards agenda-setting (Stringer, 2008).

Taking an integrated landscape approach (Figure 2.1), LDN seeks to maintain and improve the quality of land and its capacity to supply ecosystem services that can support human well-being for current and future generations (UNCCD Secretariat, 2013), placing food production at the centre of its attention. These goals

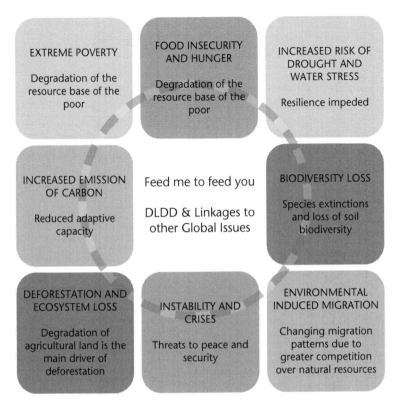

**FIGURE 2.1** The benefits of achieving a transition to a land degradation neutral world via an integrated landscape approach (from UNCCD Secretariat, 2013)

are also shared by the FAO's "Global Soil Partnership", which arose from discussions around the Millennium Development Goals. The Global Soil Partnership aims to "create a unified and recognized voice for soils through coordination and partnership, to avoid fragmentation of efforts and wastage of resources"[3].

Operationalizing LDN is not, however, straightforward. Chasek *et al.*, (2015) propose four key steps, each with their challenges: (i) scoping the scale and domain of LDN; (ii) mapping degradation; (iii) prescribing relevant sustainable land management practices; and (iv) monitoring and evaluation. Each of these is explored in more detail on the next page.

While the neutrality of land degradation needs to occur at the global scale, it requires actions to be taken at the local scale. It is the aggregate effect of local actions that will allow LDN to be achieved. Chasek *et al.* (2015) propose that an LDN domain needs to be either thematic (linked to particular land cover, e.g. rangelands, forests, etc.) or geographic (for example, a watershed or a district). This step equates to setting system boundaries, in preparation for mapping the status of the land (i.e. whether it is already degraded and abandoned, so a possible site for restoration, or already degraded and still being used, so a possible site for the application of measures that can help to reduce degradation). It is also useful to map areas that are at risk of future degradation, taking into account future climate changes too. To undertake the necessary mapping requires a clear definition of land degradation within the theme or domain of focus, and the selection of locally relevant indicators that target a range of different types of ecosystem service. Methodologies then need to be developed in order to measure and assess the processes and states of degradation within the system and to establish baselines. Decisions are also needed on where the tipping points might be such that land becomes degraded or is considered non-degraded and therefore restored. Actions to reorient land management practices towards more sustainable land management trajectories then need to be identified, so that they specifically target both the drivers and symptoms of degradation that have been identified. Different practices may be required for restored land compared to land on which preventative measures can be applied. The final step is that of monitoring and assessment in order to evaluate the progress of the management practices to deliver LDN within the domain of focus. This should not only take place at the end of projects, but throughout (Reed *et al.*, 2011), and requires independent verification.

For the process outlined above to be enacted requires an enabling environment that provides sufficient financial resources, a programme of awareness raising, motivation and empowerment, particularly of weak institutions and marginalized groups. It will also require closure of the loop between bottom-up local actions and top-down implementation of a global LDN world, such that the aggregation of local-scale efforts is recognized within the international LDN framework.

## 2.6 Synthesis

This chapter has provided the international policy context for both climate change and land degradation, focusing on the UNFCCC, UNCCD and UNCBD. One of the most distinctive characteristics of the UNCCD is its consideration of local knowledge via civil society organizations and other mechanisms. However, the UNCCD has only recently developed a process for knowledge and research to inform its decision making through the creation of its SPI. The integration of scientific knowledge has increasingly brought the systemic links between issues tackled by each Convention into focus. As a result, increased attention has been paid to the overlaps and synergies between the Rio Conventions, and wider cross-cutting issues, such as water security, conflict and migration. A number of mechanisms have now been developed to avoid overlaps and harness synergies between each of the Rio Conventions.

Part of this trend towards more "joined-up thinking" between the Conventions has been the more recent pursuit of LDN. LDN offers a framework for action that seeks to reduce degradation and scale up restoration activities from community to landscape scales. LDN accepts that further degradation is inevitable, due to pressures from growing populations and climate change, but emphasizes the restoration of degraded areas through an integrated landscape approach, to provide ecosystem services (including climate regulation), biodiversity and food and water security, for the well-being of current and future generations.

This policy context sets the international stage upon which the drama will play out between climate change and land degradation over the coming years. However, before we lift the curtain, it is important to go backstage, and to examine some conceptual and methodological frameworks that can help us understand the drama that is likely to unfold. The frameworks outlined in the following chapter will help us to interpret the likely implications for human populations and ecosystems around the world, and help us to identify actions that may enable us to avoid or adapt to some of the most concerning interactions that may take place between climate change and land degradation.

## Notes

1 In addition to these developments, in 2013 the Intergovernmental Technical Panel on Soils (ITPS) was established to provide scientific and technical advice to FAO's Global Soil Partnership: www.fao.org/globalsoilpartnership/intergovernmental-technical-panel-on-soils/en/
2 For more information, visit: www.ipbes.net/index.php/work-programme/objective-3/45-work-programme/459-deliverable-3bi.html
3 www.fao.org/globalsoilpartnership/en/

## References

Barkemeyer, R., Stringer, L.C., Hollins, J.A., Josephi, F. 2015. Corporate reporting on solutions to wicked problems: sustainable land management in the mining sector. *Environmental Science and Policy* 48: 196–209.

Bauer, S., Stringer, L.C. 2009. The role of science in the global governance of desertification. *Journal of Environment and Development* 18: 248–267.

Chasek, P., Safriel, U., Shikongo, S., Fuhrman, V.F. 2015. Operationalizing zero net land degradation: the next stage in international efforts to combat desertification? *Journal of Arid Environments* 112: 5–13.

FAO. 1993. Role of forestry in combating desertification. *FAO Conservation Guides* 21.

Fischer, H., Behrens, M., Bock, M., Richter, U., Schmitt, J., Loulergue, L., Chappellaz, J.A., Spahni, R., Blunier, T., Leuenberger, M., Stocker, T. F. 2008. Changing boreal methane sources and constant biomass burning during the last termination. *Nature* 452: 864–867.

Gregory, P.J., Ingram, J. S., Brklacich, M. 2005. Climate change and food security. *Philosophical Transactions of the Royal Society B: Biological Sciences* 360: 2139–2148.

IPCC. 2016. History. Available online at: https://www.ipcc.ch/organization/organization_history.shtml [accessed 23 February 2016].

MA (Millennium Ecosystem Assessment). 2005. *Ecosystems and Human Well-being: Current state and trends assessment.* Island Press: Washington, D.C., USA.

Paavola, J., Adger, W.N., Huq, S. 2006. Multifaceted justice in adaptation to climate change. In: Adger, W.N., Paavola, J., Huq, S., Mace M.J. (eds). *Fairness in Adaptation to Climate Change.* The MIT Press, pp. 263–277.

Parry, M., Rosenzweig, C., Iglesias, A., Fischer, G., Livermore, M. 1999. Climate change and world food security: a new assessment. *Global Environmental Change – Human and Policy Dimensions* 9: S51–S67.

Parry, M., Rosenzweig, C., Livermore, M. 2005. Climate change, and risk global food supply of hunger. *Philosophical Transactions of the Royal Society B – Biological Sciences* 360: 2125–2138.

Reed, M.S., Buenemann, M., Atlhopheng, J., Akhtar-Schuster, M., Bachmann, F., Bastin, G., Bigas, H., Chanda, R., Dougill, A.J., Essahli, W., Evely, A.C., Fleskens, L., Geeson, N., Glass, J.H., Hessel, R., Holden, J., Ioris, A., Kruger, B., Liniger, H.P., Mphinyane, W., Nainggolan, D., Perkins, J., Raymond, C.M., Ritsema, C.J., Schwilch, G., Sebego, R., Seely, M., Stringer, L.C., Thomas, R., Twomlow, S., Verzandvoort, S. 2011. Cross-scale monitoring and assessment of land degradation and sustainable land management: a methodological framework for knowledge management. *Land Degradation & Development* 22: 261–271.

Stringer, L.C. 2008. Reviewing the International Year of Deserts and Desertification 2006: what contribution towards combating global desertification and implementing the United Nations convention to combat desertification? *Journal of Arid Environments* 72: 2065–2074.

Stringer, L.C., Scrieciu, S.S., Reed, M.S. 2009. Biodiversity, land degradation, and climate change: Participatory planning in Romania. *Applied Geography* 29: 77–90.

Tal, A. 2015. The implications of environmental trading mechanisms on a future Zero Net Land Degradation protocol. *Journal of Arid Environments* 112: 25–32.

Thomas, R.J. 2008. 10th Anniversary review: addressing land degradation and climate change in dryland agroecosystems through sustainable land management. *Journal of Environmental Monitoring* 10: 595–603.

UNCCD Secretariat. 2013. A stronger UNCCD for a land-degradation neutral world. Issue Brief, Bonn, Germany.

UNCCD. 1994. *United Nations Convention to Combat Desertification in Those Countries Experiencing Serious Drought and/or Desertification Particularly in Africa: Text with annexes.* UNEP: Nairobi.

UNCCD. 2008. *The 10-year strategic plan and framework to enhance the implementation of the Convention (2008–2018).* Secretariat of the UNCCD, Bonn, Germany. Available online at: www.unccd.int/Lists/SiteDocumentLibrary/10YearStrategy/Strategy-leaflet-eng.pdf [accessed 23 February 2016].

UNCCD. 2015. Benefits and responsibilities of Parties to the Convention. Available online at: www.unccd.int/en/about-the-convention/the-convention/benefits/Pages/default.aspx

UNFCCC (United Nations Framework Convention on Climate Change). 1992. Text with annexes. Available online at: http://unfccc.int/key_documents/the_convention/items/2853.php

UNGA. 2004. Official records of the general assembly, fifty-eighth session. Supplement No. 25 (A/58/25), Annex.

UNGA. 2012. Resolution adopted by the general assembly on 27 July 2012, sixty-sixth session. (A/66/L.56), Annex.

Winslow, M.D., Vogt, J.V., Thomas, R.J., Sommer, S., Martius, C., Akhtar-Schuster, M. 2011. Science for improving the monitoring and assessment of dryland degradation. *Land Degradation & Development* 22: 145–149.

# 3

# CONCEPTUAL AND
# METHODOLOGICAL FRAMEWORKS

The interactions between climate change and land degradation in regions affected by DLDD are complex and in large part unknown. This is largely because climate change and land degradation comprise many different processes operating over different temporal and spatial scales. It is challenging enough to predict how either of these processes may play out in the future on their own, let alone consider the way they may act together. Different parts of the world are, and will be, affected by the combined processes and impacts of land degradation and climate change in different ways. Understanding which areas will fare better or worse than others requires the use of concepts and methodologies that enable us to identify the drivers of land degradation and climate change. Once we understand what is driving the problems, it is possible to assess which areas and systems are most exposed and sensitive to them.

This chapter sets out some of the climate changes that are already happening and illustrates the diverse impacts that they are having in locations across the world. It then reviews some of the conceptual frameworks that allow us to better understand the drivers, interactions and impacts within particular human and environmental systems. These frameworks are then synthesized and operationalized via the presentation of a methodological approach to assess the vulnerability of social-ecological systems to land degradation and climate change. Our frameworks are used to structure the rest of the book, providing the conceptual basis for identifying key vulnerabilities (Chapters 4 and 5) and responses (Chapter 6) to the interactive effects of climate change and land degradation, and the methods to monitor and assess these changes (Chapters 7 and 8).

## 3.1 Climate change impacts across the world

The evidence that we are in an era of unprecedented climate change is building. There was a global average increase in land and ocean temperatures of 0.85°C between 1880 and 2012, with each of the last three decades successively warmer than any of the preceding decades since 1850 (IPCC, 2013). There is nevertheless, little evidence for long-term drying trends so far, and predictions of future change in precipitation have relatively low confidence. However, there have been more heat waves and heavy rainfall events in some parts of the world, with some of these events being the most extreme for centuries.

According to IPCC (2013), increases in the severity and duration of droughts are likely by the second half of the twenty-first century, but again, such predictions have low confidence for the first half of the century.

According to the latest assessment from the Intergovernmental Panel on Climate Change (IPCC, 2014), strong and widespread impacts have now been observed across the world on natural systems (Figure 3.1), with impacts on hydrological

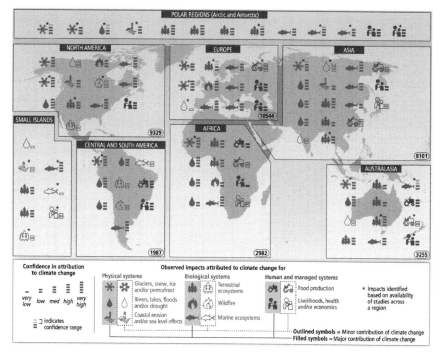

**FIGURE 3.1** Observed impacts of climate change based on scientific literature published 2007–2011. Symbols indicate types of impact, the relative contribution of climate change (major or minor) to the observed impact and confidence in attribution. Numbers in ovals indicate regional totals of climate change publications from 2001–2010, based on the Scopus bibliographic database for publications in English with individual countries mentioned in title, abstract or key words (these numbers provide an overall measure of the available scientific literature on climate change across regions; they do not indicate the number of publications supporting attribution of climate change impacts in each region)

systems and ecosystems having a number of knock-on effects on food production and livelihoods. However, interactions between climate change and land degradation will be felt very differently around the world. This is partly because the climate is likely to change in different ways in different regions; there will be different levels of warming, and some areas will become drier while others become wetter.

According to IPCC (2014), in Africa, there is already a reduction in the amount of water being discharged from West African rivers, creating water shortages for human consumption and agriculture. There has been a warming of water in the Great Lakes and Lake Kariba, which has resulted in the decline in productivity of fisheries. The density of trees has decreased significantly in recent years in the western Sahel and semi-arid Morocco, beyond the sorts of levels that could reasonably be attributed to land use change. Declines are especially notable for fruit trees, which have a direct impact on the livelihoods of local populations who have traditionally used these fruits. Shifts in the areas or range of many species of plants and animals have been observed, particularly in southern Africa, and there has been an increase in the number of wildfires on Mount Kilimanjaro. Furthermore, as temperatures have risen, malaria has spread in the Kenyan highlands.

In Europe, there is evidence of major changes in the functioning of ecosystems due to climate change, including earlier greening, leaf emergence, and fruiting in temperate and boreal trees. A number of alien plant species are spreading across the continent as conditions become more favourable due to climate change. Migratory birds have been arriving earlier in Europe since 1970, and this trend is continuing. There is a steady northward and depth shift in the distribution of many fish species in European seas, and a spread of warm-water species into the Mediterranean. Forest fires are increasing in severity and extent, especially in drier areas, such as Portugal and Greece. Despite significantly higher levels of development in Europe, compared to Africa, human impacts have been considerable. Heat waves are causing increasing mortality (see Box 3.1), and the livelihoods of groups like the Sámi people in northern Europe have been significantly compromised due to the thawing and freezing of ice as a result of climate change. This is problematic because it creates a crust of hard ice over the lichens that the reindeer feed on, preventing them from being able to access a key food source. Impacts on agriculture have varied, with wheat yields in some countries stagnating in recent decades, despite improvements in technology, and positive impacts on yields for some other crops, mainly in northern Europe. The spread of bluetongue virus in sheep, and of ticks across parts of Europe, has also been linked to climate change (IPCC, 2014).

In Asia, there is evidence of permafrost melt in Siberia, Central Asia, and the Tibetan Plateau, and bush encroachment into the Siberian tundra. Mountain glaciers are shrinking across most of Asia, leading to increased flow in several rivers. However, the quality of surface waters has declined across Asia. There is evidence of a reduction in the availability of water in many Chinese rivers, beyond changes that might be expected due to land use shifts, and a reduction in soil moisture was observed in north-central and northeast China over the period 1950–2006. This has led to negative impacts on wheat and maize yields. Beyond this, declining

## BOX 3.1: THE 2003 EUROPEAN HEATWAVE

Europe is generally considered to have high levels of adaptive capacity when it comes to dealing with climate change. However, the heatwave that occurred during August 2003 was the warmest period of extreme heat for 500 years and highlighted how unprepared much of the continent was for such extreme events. This is an important event to learn from because projections suggest that summers as hot as that of 2003 could be occurring every other year by the year 2050 as a result of climate change.

In the 2003 case, a high pressure system had settled over most of Western Europe, bringing dry tropical continental air with little cloud. Temperature records were broken in several parts of Europe, while rainfall was lower than expected during June, July and August. The heatwave resulted in a number of biophysical and human costs. The high temperatures led to high evaporation and low river flows. The flow of the River Danube in Serbia reached its lowest level in 100 years. Previously submerged bombs and tanks from the Second World War caused a hazard to people swimming in the river, while the output of hydro-electric power plants substantially reduced, creating energy generation challenges. In Germany, two nuclear power plants were closed due to lack of water needed to assist with cooling. In Portugal, forest fires decimated more than 200,000 ha of forest, leading to the erosion of millions of tonnes of topsoil due to the bare ground this created. Water quality also declined, where much of the eroded soil ended up in water bodies used as drinking water sources.

More than 20,000 people died – 15,000 of these in France due to the heat. The sheer number of deaths required makeshift mortuaries to be set up in refrigeration lorries, as existing infrastructure was unable to cope. Cases of heatstroke, dehydration and sunburn increased, while increased numbers of people engaging in tourism and cooling off in rivers and lakes led drowning cases to rise. Up to a third of the UK deaths during the heatwave were linked to breathing problems related to poor air quality. Water supplies were affected across Europe and in some countries, such as Croatia, hosepipe use was banned. Many farm animals died and crop yields declined, resulting in higher food prices. Reports suggest that the heatwave cost European farming 13.1 billion euros. In some locations, road surfaces melted, while low river levels prevented sailing. In the UK, trains were subject to speed restrictions when temperatures rose above 30 degrees to help reduce the risk of derailment where railway lines were starting to buckle in the heat. The French government called on EU aid to help deal with the impacts, while public information initiatives on how to deal with the heat were stepped up. Since 2003, the French government has enhanced its early warning systems, particularly in relation to surveillance and alerts for particularly vulnerable groups such as the elderly.

Modfied from www.metoffice.gov.uk/learning/learn-about-the-weather/weather-phenomena/case-studies/heatwave wheat yields attributed to climate change have been observed across Asia. Many plant and animal species are now found further north and at higher altitudes than ever before, particularly in the north of Asia, for example, Siberian larch forests have seen invasion by pine and spruce in recent decades. Similarly, there has been a range extension of corals in the East China Sea and western Pacific, while coral reefs are declining in tropical Asian waters (IPCC, 2014).

There have been similar impacts across Australasia. Late season snow depth has been decreasing in alpine areas of Australia since records began in 1957, and there has been a substantial reduction in ice and glacier ice volume in New Zealand. The amount of water discharged by rivers in southwestern Australia has declined and there has been an intensification of hydrological drought due to regional warming in southeast Australia. Changes have been observed in the genetics, growth, distribution and phenology of many plants and animals (especially birds and butterflies) in Australia, beyond the sorts of fluctuations that might be expected due to variability in local climates, land use and pollution. The range of many marine species has shifted southwards in recent decades, and there is evidence of increased coral bleaching in the Great Barrier Reef. Climate change has led to a shift in human mortality from winter towards summer, and a relocation or diversification of agricultural activities across Australia, in response to changes in climate (IPCC, 2014).

In North America, glaciers have been shrinking, with a decreasing amount of water held in spring snowpacks and earlier peak flows in snow-dominated rivers in western North America. There is evidence of changes in the phenology and distribution of many plants and animals, northwards and to higher altitudes. Similarly, the distribution of many Atlantic fish populations has shifted northwards and the migration patterns and survival of salmon in northeast Pacific waters has changed in response to climate change. Wildfires are now more frequent in subarctic conifer forests and tundra, and regional increases in tree mortality and insect infestations have been observed in many North American forests. These changes have had significant impacts on the livelihoods of indigenous groups in the Canadian Arctic (IPCC, 2014).

In Central and South America, Andean glaciers have been shrinking and there are changes in discharge patterns, including extreme flows in the Amazon and rivers in the Western Andes. There is evidence of increased tree mortality and forest fires in the Amazon and mangrove degradation on the north coast of South America. These changes are having a range of effects on human populations. For example, increases in agricultural yields and an expansion of agricultural areas in southeastern South America have been attributed to climate change. On the other hand, the livelihoods of indigenous Aymara farmers in Bolivia are particularly vulnerable to climate change due to water shortages.

Apart from being threatened by sea level rise (leading to increased flooding and erosion and saline intrusion into groundwater), small island states are experiencing a decline in their coastal fisheries due to direct effects of climate change and indirect effects of increased coral reef bleaching. Increased water scarcity has been observed in Jamaica, beyond changes that could be expected due to increases in water usage (IPCC, 2014).

## 3.2 Conceptual frameworks

There are many different ways of conceptualizing the links between the climate changes outlined in the previous section, and land degradation, as well as how these interacting processes might influence the environment and human well-being. There are also many ways of interpreting the vulnerability of ecosystems and human well-being to these drivers of change (Reed *et al.*, 2013a). Understanding land degradation and climate change requires an emphasis to be placed on unraveling relationships, trade-offs and feedbacks. This needs to happen alongside recognition that lots of interacting factors (both biophysical and socio-economic) operate at different rates over different time frames to result in particular outcomes for particular groups of people in particular places. It is now increasingly appreciated that the health and quality of the environment directly and indirectly affect the health and well-being of human populations. Consequently, relationships are recognized between environmental degradation and the economy, health, migration, conflict, education, disaster risk reduction and development. To understand how these aspects all fit together requires a focus on systems. In this book we use two linked systems-based conceptual frameworks to explain how land degradation and climate change relate to human well-being through effects on ecosystem processes and services, and then to explain how we can assess the vulnerability or resilience of these ecosystems and their human populations to land degradation and climate change.

The first conceptual framework is shown in Figure 3.2. This framework is used by the IPBES to show the main elements and relationships for the conservation and sustainable use of biodiversity and ecosystem services, human well-being and sustainable development. Different forms of land degradation and climate change are conceptualized as drivers. These drivers either directly or indirectly (depicted by dotted arrows) influence human well-being, through their effects on biodiversity and ecosystem processes, and the provision of ecosystem services. It shows how drivers of change such as land degradation and climate change are part of a social-ecological system that can be influenced by institutions, governance, other indirect drivers of change, e.g. cultural factors that might enable or present barriers to tackling climate change and land degradation.

Similar conceptualizations in other knowledge systems include "living in harmony with nature" and "Mother Earth", among others. In the main panel, delimited in grey, "nature", "nature's benefits to people" and "good quality of life" (indicated as black headlines) are inclusive of all these world views.

Although stocks of natural capital are not identified explicitly in the framework, they implicitly form part of "nature", and in combination with stocks of physical, human, social and financial capital can provide flows of ecosystem goods and services that can support human well-being. The framework explicitly considers how these different components of social-ecological systems are linked and change over different spatial and temporal scales. Although climate change is a global process, this book shows how it interacts with different forms of land

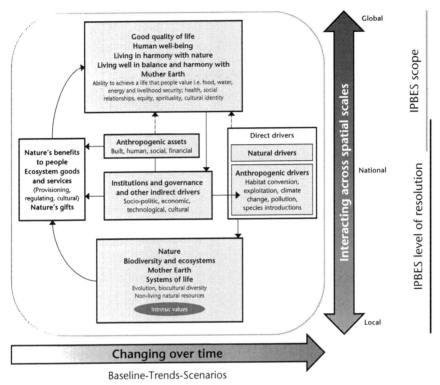

**FIGURE 3.2** Conceptual framework used by the Intergovernmental Panel on Biodiversity and Ecosystem Services, showing the main elements and relationships for the conservation and sustainable use of biodiversity and ecosystem services, human well-being and sustainable development. Text in light grey denotes the concepts of science, while text in dark grey denotes those of other knowledge systems. Solid arrows in the main panel denote influence between elements; the dotted arrows denote links that are acknowledged as important, but are not the main focus of the Platform. The thick arrows below and to the right of the central panel indicate different scales of time and space, respectively (From IPBES Secretariat, 2014)

degradation at different spatial scales, affecting ecosystem processes operating from micro- to macro-scales, which then impact upon the provision of ecosystem services and people's livelihoods. It considers how baseline levels of natural capital are likely to change over time to generate trends under climate change scenarios, in the presence of land degradation. It also considers how different anthropogenic assets may be used to mitigate or adapt to the effects of climate change and land degradation.

However, it is important to recognize that impacts on human populations are not necessarily inevitable if they are exposed to climate change and land degradation. Some ecosystems, ecological processes, ecosystem functions, ecosystem services and human populations may be more vulnerable than others to drivers of change. It is therefore necessary to couple the conceptual framework in Figure 3.2 with an understanding of the factors that increase or decrease the vulnerability of different social-ecological systems to land degradation and climate change.

## 3.3 Understanding the vulnerability of ecosystems and populations in areas affected by DLDD

To understand the likely effects of land degradation and climate change on any given ecosystem or human population, it is necessary to understand how vulnerable they are to these drivers of change. The concept of vulnerability usually relates to the degree to which a human social and/or ecological system will be affected by some form of hazard (Turner *et al.*, 2003). Hazards can take the form of major spikes in some kind of pressure (e.g. extreme weather events including drought), or stresses, which are continuous, slowly increasing pressures (such as soil degradation). In addition, some spikes may have a cumulative effect, especially when added to underlying stresses. Hazards can arise from both within and outside the ecosystems, communities and regions affected by DLDD.

Although there are varying interpretations, three factors are regularly discussed in the literature with regard to vulnerability, and these form the basis for the second conceptual framework, which is presented in Figure 3.3. By understanding each of

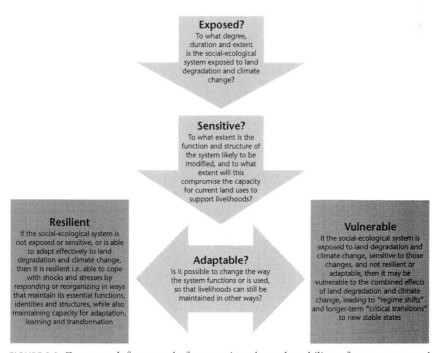

**FIGURE 3.3** Conceptual framework for assessing the vulnerability of ecosystems and human populations to land degradation and climate change

these factors, it is possible to identify how vulnerable or resilient different ecosystems and human populations might be to the joint effects of land degradation and climate change. The next step is to work out how to reduce this vulnerability. Figure 3.3 summarizes the three factors that determine the vulnerability or resilience of ecosystems and populations to land degradation and climate change. Broadly, the components in Figure 3.3 correspond to the vulnerability framework developed by the IPCC, which has been widely adopted for assessing the susceptibility of systems to the effects of climate change and other human stressors such as land degradation (Mumby *et al.*, 2014).

*Exposure* considers the degree, duration and extent to which the ecosystems and populations are exposed to land degradation and climate change. The concept of risk in its broadest sense overlaps with vulnerability in a number of ways. For the purposes of this conceptual framework, risk is defined as the probability that exposure to land degradation and climate change will lead to negative impacts on ecosystems and human populations in regions affected by DLDD, as mediated by the capacity for that system to adapt to the pressures to which it is exposed and sensitive. Exposure is sometimes incorporated in the concept of risk (e.g. IPCC, 2012).

*Sensitivity*: if the system is exposed to land degradation and climate change, then its sensitivity can be defined as the extent to which the function and structure of ecosystems are likely to be modified by the changes they are exposed to, and the extent to which this will compromise the capacity for current land uses to support livelihoods. Alternatively this can be conceptualized as the 'stability' of the system or its capacity to retain essential functions and structures in the face of pressures from land degradation and climate change, and its capacity to deliver essential services.

*Adaptability*: if the system is exposed and sensitive to the effects of land degradation and climate change (e.g. increased incidence and severity of droughts), then it is necessary to assess the adaptive capacity of the system, i.e. the extent to which it is possible to change the way the system functions or is used, so that livelihoods can still be maintained in other ways. Adaptation may take the form of: coping (by enacting short-term, immediate responses to reduce risk from climate variability and drought to livelihoods); adjustment (more deliberate planned change, representing adaptation to longer-term climate change and land degradation); and/ or transformation (fundamental changes to either system function or political economic structures, often involving behavioural changes, and leading to the establishment of new long-term social-ecological states) (Folke *et al.*, 2010; Béné *et al.*, 2012; Keck and Sakdapolrak, 2013; Stringer *et al.*, in press).

It is important to note that many apparent adaptations to climate change and land degradation may actually be maladaptive, if they are not sustainable or if they increase vulnerability. Vulnerability can be increased by worsening the effects of land degradation and climate change on other ecosystems and populations, or by locking people into particular livelihood trajectories that compromise their future adaptive capacity. It is also important to recognize that adaptations are often context and scale dependent; an adaptation for one community or system at one scale may or not be useful in another community or system at a different scale (Stringer *et al.*, in press).

If the system of focus is exposed, sensitive and unable to adapt effectively to the effects of land degradation and climate change, then it will not be able to maintain its essential functions, identities and structures or its ability to adapt to future changes, and it will become *vulnerable* to land degradation and climate change. This may lead to significant changes in the social-ecological system, sometimes referred to as "regime shifts" (Scheffer *et al.*, 2001; Carpenter, 2003) and "critical transitions" when these shifts lead to new long-term stable states (Scheffer, 2009).

Systems that are able to adapt effectively to the effects of land degradation and climate change may in many cases be *resilient*. Resilience considers the "ability of a social-ecological system to cope with shocks and stresses by responding or reorganizing in ways that maintain its essential functions, identities and structures, while also maintaining capacity for adaptation, learning and transformation" (adapted from Arctic Council, 2013, cited in IPCC, 2014). In the context of land degradation and climate change, this is "general resilience" – it considers the resilience of whole systems rather than the "specified resilience" of individual system components (Folke *et al.*, 2010; Stringer *et al.*, in press). If general resilience is low, then so is the overall resilience of the system.

Particular groups are often identified as in need of special attention when it comes to vulnerability assessment and the development of solutions to build resilience. In the context of DLDD, the poor are one such group. The drylands alone are inhabited by more than 2.5 billion people, 16 per cent of whom live in chronic poverty. The poor are largely dependent upon the land for their survival, which leaves them both exposed and sensitive to climate impacts and degradation, particularly because they lack the capacity and assets to be able to respond to the shocks and stresses that affect their livelihoods. The poor are also marginalized when it comes to both participating in formal institutional decision making and gaining the recognition that affords them the opportunity to participate. Other groups that are often considered particularly vulnerable are women (largely due to the triple burden they face regarding the need to run a household, engage in paid work as well as taking responsibility for childcare), children and the elderly (who often lack the human capital necessary to adequately manage shocks and stresses). Climate change and land degradation amplify the day-to-day challenges these groups face perhaps more so than for other groups. As a result, many development programmes tasked with building resilience and reducing vulnerability place special focus on these groups.

## 3.4 Methodological framework

There are many approaches to assessing vulnerability and responding to land degradation and climate change (Reed *et al.*, 2013a). Following the IPBES framework in Figure 3.2, it is important that any methodological frameworks for vulnerability assessment should consider how impacts on human well-being are likely to be mediated by effects on ecosystem processes, ecosystem functions and the provision of ecosystem services to these populations. Methods for assessing vulnerability can be qualitative or quantitative, and can be applied from local to international scales.

Increasingly mixed method approaches are being used to assess vulnerability, combining qualitative and quantitative methods from a range of disciplinary traditions. The assumptions that sit behind these different methods can be widely varied, and without careful integration, there is a danger that methods from different traditions may provide contradictory findings that do not usefully inform policy or practice. This is because perceptions of what constitutes valid knowledge and evidence differ between research traditions (known in social science as "epistemology").

In some disciplines, knowledge may be seen as black and white; either something is universally true or it is false. Researchers who hold this "positivist" or "reductionist" view of knowledge are likely to create and test hypotheses to deduct principles and theory, and break problems and dryland systems into their constituent parts to try and piece together solutions. Researchers working in this way tend to give greater weight to formal and explicit knowledge and quantitative, statistically robust and objective evidence, rather than more informal, implicit or tacit knowledge based on qualitative data, which may be perceived as being more subjective.

In other disciplines, knowledge may be seen in shades of grey along a spectrum: rather than seeking universal truths and solutions, subjective realities and context-dependent options are sought. Researchers who hold this more post-modern view of knowledge generate theory inductively, grounded in bodies of text or conversation based on the narratives of the people affected by the challenges they are researching. They are more likely to focus on connections, feedbacks, uncertainties and alternative conflicting accounts, viewing the system as a whole wherever possible, rather than focusing on its constituent parts. They are more likely to give equal weight to informal and implicit or tacit knowledge and qualitative data that can explore different subjective perspectives. In response to critiques that evidence based on the latter approach is less robust than the former approach, qualitative data sources will typically be checked against each other or "triangulated". So for example, different oral histories covering the same period may be checked against each other for consistency; they may be checked against documentary evidence about key events; and they may be discussed in group settings. Triangulation is important in mixed methods research too, particularly in resolving any conflicts between data collected using different methods.

When designed appropriately, triangulation in mixed methods research can provide a highly complementary picture of the vulnerability of a system to climate change and land degradation. Rather than providing contradictory accounts, quantitative secondary data may be analyzed to provide a baseline understanding of the context, including climatic trends and indicators of the severity and extent of land degradation. In this context, qualitative analyses may then be able to help interpret the significance of trends detected from quantitative data. For example, the extent to which a local community is vulnerable to the effects of land degradation and climate change will depend on the effects that these processes have on the ability of community members to sustain their livelihoods. Qualitative data, for example, based on interviews with land managers, might provide important information about the capacity of local communities to adapt to changes in natural resource availability arising from land degradation and climate change. Findings gained through these approaches may identify that the people involved in the study are significantly less vulnerable to these processes than a community that does not have particular adaptive capacities and options. For example, as described in Chapter 1, Warren (2002) questions whether a community whose land is experiencing severe erosion in West Africa could be described as vulnerable to land degradation and climate change, if they are able to consistently maintain yields of agricultural outputs from that land.

The methodological framework in Figure 3.4 is designed to operationalize the conceptual frameworks outlined in Figures 3.2 and 3.3, and can be used following a mixed method approach.

Figure 3.4 shows how each step in the methodological framework relates to the conceptual framework presented in Figure 3.3. It is represented here as a circular

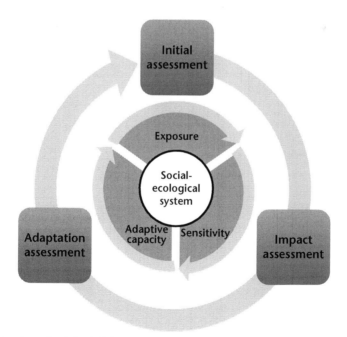

FIGURE 3.4 A methodological framework (outer circle) for assessing the vulnerability of ecosystems and human populations in regions affected by DLDD to the combined effects of climate change and land degradation (segmented middle circle, based on conceptual framework in Figure 3.2; inner circle based on conceptual framework in Figure 3.3)

ocess, to emphasize that vulnerability assessments provide a snap-shot in time, and as systems are exposed to new threats, or change their sensitivity or adaptive capacity, assessments will need to be revisited. It consists of the following steps, each of which is explored in greater depth in later sections of this book:

1.  *Initial assessment:* evaluation of the degree to which the stocks of natural capital and ecosystem processes ("nature" in Figure 3.2), and flows of ecosystem services are exposed to drivers of change – in this case, climate change and land degradation (the upper arrow in the conceptual framework in Figure 3.3: exposure). The exposure of a system to climate change can be assessed historically from climate records, while future climates may be projected using the sorts of predictive models described in the next section. There is an extensive literature on methods for assessing exposure to land degradation (whether actual or considering the probability of degradation). Broadly, these can be classified as methods for: (i) direct measurement (e.g. of soil fertility and productivity); (ii) indirect measurement via indicators (e.g. soil erosion features and vegetation cover); and (iii) indirect measurement and projections via process-based computational models, which typically combine a range of indicators and would be calibrated and validated via direct measurements. At local scales such assessments may combine qualitative social science methods (e.g. semi-structured interviews, oral histories and ethnographic methods) with quantitative methods based on indicators (e.g. GIS mapping or process-based modelling of the effects of land degradation and climate change on land cover, populations of animals and plants, and livestock populations). At regional and international scales, assessments may be based on expert opinion (e.g. the Global Assessment of Soil Degradation (GLASOD) upon which UNEP's (1997) World Atlas of Desertification was based), or process-based models, (e.g. models of future agricultural yields or forest cover based on projections from Global Circulation Models).

2.  *Impact assessment:* this refers to an evaluation of the sensitivity to climate change and land degradation (the second arrow in Figure 3.3) of each of the core components of the social-ecological system described in Figure 3.2, and hence an assessment of likely impacts on human well-being in the absence of adaptation. To understand the sensitivity of ecosystems and populations in regions affected by DLDD to the combined effects of climate change and land degradation, it would be necessary to know the extent to which changes in air and soil temperature, precipitation (total amount, intensity/erosivity and patterns), humidity, atmospheric $CO_2$ concentrations and evapotranspiration rates are likely to affect land degradation processes, and so compromise the supply of ecosystem services and the livelihoods and human well-being that depend on them. Many of the same models used to assess degradation severity, extent and/or risk may be used or adapted to assess these links. However, given the complexity of the relationships and feedbacks between climate change and land degradation and the complexity of the systems they are affecting, models can

only provide an approximate assessment of plausible outcomes. Given the approximate nature of model outputs in such complex social-ecological systems, there are also strong arguments for including evidence based on locally held knowledge of how these systems work, which can often provide highly complementary information. Locally held and scientific knowledge along with qualitative, approximate and incomplete information may be integrated using techniques such as mediated or participatory modelling, dynamic systems modelling and Bayesian Belief Networks. Many of these methods are discussed further in later chapters.

3.  *Adaptation assessment:* this considers the potential and feasibility of adaptive capacity to reduce the sensitivity of the system to the changes it is likely to be exposed to, and provides specific recommendations to planners and policy-makers. Using social science methods, it may also be possible to catalogue how local communities have adapted to previous changes in the productive potential of the land or climate variability, and to provide insights into potential future adaptations to land degradation and climate change. Process-based models may be used in a number of different ways to gain further insight into the future pressures likely to arise from land degradation and climate change, and to evaluate and refine adaptive options based on local innovations and scientific research in light of model outputs (Prell *et al.*, 2007; Reed *et al.*, 2013b). However there are a range of limitations and uncertainties associated with these techniques.

Initial impacts and adaptation assessments require trans-disciplinary approaches that combine knowledge from scientific and other stakeholders, as well as combining experimental (field and laboratory) research, modelling and multi-level stakeholder engagement. Multi-level stakeholder engagement is especially important in order to elicit local knowledge, to select feasible tailor-made solutions that fit to local socio-economic and cultural conditions, and to establish buy-in from high level policy-makers and implementers (de Vente *et al.*, in press).

## 3.5 Synthesis

This chapter has reviewed some of the key conceptual and methodological approaches for assessing the vulnerability of social-ecological systems to the combined effects of land degradation and climate change. Based on this, a conceptual framework has been synthesized and combined with the IPBES conceptual framework and insights from elsewhere. These concepts have then been used to develop a methodological framework that can be applied to assess the vulnerability or resilience of ecosystems and populations to land degradation and climate change. The methodological framework may be used by policy-makers, researchers, practitioners and other stakeholders in order to: assess the extent to which social-ecological systems are exposed to land degradation and climate change via direct or indirect measurements and projections; and evaluate sensitivity and capacity to adapt to those changes by combining insights from biophysical and social assessments, including locally held knowledge. It is important to recognize the potential for maladaptation, and that there may be limits to adaptation, given the speed with which land degradation and climate change may occur in some contexts.

The next two chapters apply the methodological framework developed in this chapter to consider current and likely future exposures and sensitivities of ecosystem services to interactions between climate change and land degradation. In doing this, they provide an initial assessment of the vulnerabilities of people and ecosystems in regions affected by DLDD. This is followed by a consideration of responses to potential vulnerabilities in Chapter 6, including adaptation, mitigation and rehabilitation.

# References

Béné, C., Wood, R.G., Newsham, A. Davies, M. 2012. Resilience: new utopia or new tyranny? Reflection about the potentials and limits of the concept of resilience in relation to vulnerability reduction programmes. *IDS Working Papers* 2012: 1–61.

Carpenter, S.R. 2003. Regime shifts in lake ecosystems: pattern and variation. *Excellence in Ecology Series*, Volume 15. Ecology Institute: Oldendorf/Luhe, Germany.

De Vente, J., Reed, M.S., Stringer, L.C., Valente, S., Newig, J. (in press). How does the context and design of participatory decision-making processes affect their outcomes? Evidence from sustainable land management in global drylands. *Ecology & Society*.

Folke, C., Carpenter, S.R., Walker, B., Scheffer, M., Chapin, T., Rockström, J. 2010. Resilience thinking: integrating resilience, adaptability and transformability. *Ecology and Society* 15(4): 20. Available online at: www.ecologyandsociety.org/vol15/iss4/art20/

IPBES Secretariat. 2014. Decision IPBES-2/4: Conceptual framework for the Intergovernmental Science-Policy Platform on Biodiversity and Ecosystem Services. *IPBES-2 Decisions* 2014. Intergovernmental Platform on Biodiversity and Ecosystem Services: Bonn, Germany, 9 pp.

IPCC. 2012. Managing the Risks of Extreme Events and Disasters to Advance Climate Change Adaptation. In: *A special report of Working Groups I and II of the intergovernmental panel on climate change*, Field, C.B., Barros, V., Stocker, T.F., Qin, D., Dokken, D.J., Ebi, K.L., Mastrandrea, M.D., Mach, K.J., Plattner, G.K., Allen, S.K., Tignor, M., Midgley, P.M. (eds). Cambridge University Press: Cambridge, United Kingdom and New York, NY, USA, 582 pp.

IPCC 2013. Climate Change 2013. The Physical Science Basis. In: *Contribution of Working Group I to the fifth assessment report of the intergovernmental panel on climate change*, Stocker, T.F., Qin, D., Plattner, G.K., Tignor, M., Allen, S.K., Boschung, J., Nauels, A., Xia, Y., Bex, V., Midgley, P.M. (eds). Cambridge University Press: Cambridge, United Kingdom and New York, NY, USA, 1535 pp.

IPCC. 2014. Climate Change 2014. Impacts, Adaptation, and Vulnerability. In: *Part A: global and sectoral aspects. Contribution of Working Group II to the fifth assessment report of the intergovernmental panel on climate change*, Field, C.B., Barros, V. R., Dokken, D. J., Mach, K. J., Mastrandrea, M. D., Bilir, T. E., Chatterjee, M., Ebi, K. L., Estrada, Y.O., Genova, R.C., Girma, B., Kissel, E.S., Levy, A.N., MacCracken, S., Mastrandrea, P. R., White, L. L. (eds). Cambridge University Press: Cambridge, United Kingdom and New York, NY, USA.

Keck, M., Sakdapolrak, P. 2013. What is social resilience? Lessons learned and ways forward. *Erdkunde* 67: 5–19.

Mumby, P.J., Chollett, I., Bozec, Y.-M., Wolff, N.H. 2014. Ecological resilience, robustness and vulnerability: how do these concepts benefit ecosystem management? *Current Opinion in Environmental Sustainability* 7: 22–27.

Prell, C., Hubacek, K., Reed, M.S., Burt, T.P., Holden, J., Jin, N., Quinn, C., Sendzimir, J., Termansen, M. 2007. If you have a hammer everything looks like a nail: 'traditional' versus participatory model building. *Interdisciplinary Science Reviews* 32: 1–20.

Reed, M.S., Podesta, G., Fazey, I., Beharry, N.C., Coen, R., Geeson, N., Hessel, R., Hubacek, K., Letson, D., Nainggolan, D., Prell, C., Psarra, D., Rickenbach, M.G., Schwilch, G., Stringer, L.C., Thomas, A.D. 2013a. Combining analytical frameworks to assess livelihood vulnerability to climate change and analyse adaptation options. *Ecological Economics* 94: 66–77.

Reed, M.S., Hubacek, K., Bonn, A., Burt, T.P., Holden, J., Stringer, L.C., Beharry-Borg, N., Buckmaster, S., Chapman, D., Chapman, P., Clay, G.D., Cornell, S., Dougill, A.J., Evely, A., Fraser, E.D.G., Jin, N., Irvine, B., Kirkby, M., Kunin, W., Prell, C., Quinn, C.H., Slee, W., Stagl, S., Termansen, M., Thorp, S., Worrall, F. 2013b. Anticipating and managing future trade-offs and complementarities between ecosystem services. *Ecology & Society* 18: 5

Scheffer, M., Carpenter, S., Foley, J.A., Folke, C., Walker, B. 2001. Catastrophic shifts in ecosystems. *Nature* 413: 591–596.

Scheffer, M. 2009. *Critical transitions in nature and society.* Princeton University Press: Princeton, New Jersey, USA.

Stringer, L.C., Quinn, C.H., Berman, R., Dixon, J.L. In press. Livelihood adaptation and climate variability in Africa. In: Hammett, D., Grugel, J. (eds). *Handbook of International Development.* Palgrave, UK.

Turner, B.L., Kasperson, R.E., Matson, P.A., McCarthy, J.J., Corell, R.W., Christensen, L., Eckley, N., Kasperson, J.X., Luers, A., Martello, M.L., Polsky, C., Pulsipher, A., Schiller, A. 2003. A framework for vulnerability analysis in sustainability science. *Proceedings of the National Academy of Sciences of the United States of America* 100: 8074–8079.

Warren, A.S. 2002. Land degradation is contextual. *Land Degradation and Development* 13: 449–459.

# 4

# EXPOSURE AND SENSITIVITY OF PROVISIONING ECOSYSTEM SERVICES

This chapter and the next examine the likely exposure and sensitivity of human populations and ecosystems in regions affected by DLDD to climate change and land degradation. This chapter examines the likely effects on provisioning ecosystem services and food security, and consequent effects for the sustainability of livelihoods dependent on these services. Chapter 5 then considers the effects of climate change and land degradation on the other types of ecosystem services (regulating, supporting and cultural).

## 4.1 Exposure to climate change and land degradation

Land degradation is defined as a long-term, human-induced reduction in the resource potential of the land (Section 1.1), and its ability to provide a range of ecosystem services (provisioning, regulating, supporting and cultural) (MA, 2005a; Reed *et al.*, 2015). Climate change is likely to exacerbate and accelerate land degradation and ecosystem service delivery in many regions affected by DLDD, due to heat stress, drought, changes to evapotranspiration rates and biodiversity, and new diseases and pests. Indeed, the previous chapter demonstrated that these kinds of changes are already happening. The aggregate impacts of climate change and land degradation affect yields from rain-fed agriculture and livestock production – land uses which support the livelihoods and food security of many of the world's poorest people.

Climate change encompasses various factors that stem from the modification of the atmosphere's composition, including $CO_2$ concentration, air temperature, precipitation, tropospheric $O_3$ and other environmental pollutants, UV radiation, extreme events, etc. (IPCC, 2007). To assess the vulnerability of ecosystems and populations, it is necessary to pinpoint which are the most relevant climate change factors to which regions affected by DLDD are likely to be exposed. The extent to which climate change may exacerbate land degradation may be softened by increases in primary productivity and water-use efficiency due to increased concentrations of $CO_2$ in the atmosphere, and longer growing seasons in some areas due to warmer temperatures. However, the synergistic effects of rising temperatures and changes in the hydrological cycle, which may expose large areas of the globe to more frequent and severe droughts (Dai, 2011), along with increased frequencies in extreme events (such as heatwaves and mega-droughts), outweigh and may even nullify the fertilization effects of rising $CO_2$ concentration in areas affected by DLDD (Centritto et al., 2011a). Thus, the balance of effects on land degradation between these opposing and interacting climatic drivers is unclear, and exposure to these effects is likely to differ depending on location and the land management practices that are used (MA, 2005a).

Part of this uncertainty is due to the range of different, and sometimes opposing, feedbacks between climate change and land degradation processes, notably linked to carbon sequestration and storage in soils and vegetation, and the effect of changes in vegetation cover and type on the reflectivity (or "albedo") of land. There may also be feedbacks between the effects of climate change and land degradation on biodiversity and the provision of a range of ecosystem services, which can further exacerbate land degradation. These feedbacks often involve a number of degradation processes, which may affect the biological or economic productivity and resilience of land and livelihoods, for example the way carbon feedbacks and gas exchanges are mediated by changes in wind erosion (due to increased aridity under climate change), soil sealing (through urbanization and urban sprawl) and water erosion (due to increased intensity and erosivity of rainfall under climate change). These problems need to be considered in the wider context of global change challenges arising from increasing human populations and the associated demand we place upon the natural resource base. To remain within a safe operating space for humanity, we need to increase food production using less water and land, while also reducing our GHG emissions.

The changing climate will interact in different ways with different land degradation processes and land use/land management systems. For example, climate change may reduce vegetation cover, and so increase rates of soil erosion and fertility loss under conventional tillage or intense livestock grazing. However, the same interactions may have very different consequences in different places. For example, a 10 per cent loss of vegetation cover may lead to a significant rise in wind erosion on a sandy soil where the vegetation cover is relatively low to start with and a critical threshold is crossed (e.g. Wiggs et al. (1995)), calculated a 14 per cent vegetation cover threshold in south Kgalagadi District, Botswana, under which dunes are typically activated). However, the same amount of vegetation loss may have no effect on a system that started off with much more vegetation on a more mineral or a crusted sandy soil. Similarly, the ways in which the same kinds of landscapes are managed matters. Peatland areas across the world are used and managed in different ways, resulting in different outcomes from the interactions between climate change and land degradation processes. In Belarus in the 1950s, prior to intensive agricultural exploitation and peat extraction, peatlands covered nearly three million ha of the country. Peatland drainage, deemed necessary to allow agricultural land uses, caused peat mineralization, which reduced the fertility of the soil and emitted $CO_2$. Furthermore, the drained peatlands are now more susceptible to wind erosion and fire, as well as exhibiting changed solar reflection levels, compared with intact peatlands in other (biophysically similar) locations such as remote areas of Russia.

Given the uncertainties in exposure to climate change and land degradation processes in different systems around the world, it is difficult to predict how sensitive different systems, and the services that they provide to society, are likely to be to these changes. The following sections outline what we currently know about the sensitivity of provisioning services to climate change and land degradation, with a focus on the sensitivity of agriculture, forests, fresh water provision and livelihoods in regions affected by DLDD.

## 4.2 Sensitivity of agriculture

Approximately three billion ha of the Earth's surface is considered suitable for cultivation. Excluding Antarctica, more than 75 per cent of the global land surface suffers from rainfed crop production constraints (Fischer et al., 2006). About 12 per cent of the planet is too steep to grow crops; 13 per cent is too cold, while 27 per cent is too dry. These constraints make it vital to make best possible use of the remaining cultivable land, taking into account the likely future effects of climate change. The majority of research on the effects of climate change on agroecosystems has focused on changes in yield and spatial shifts in the production potential of cropping systems (e.g. Reilly and Schimmelpfennig, 1999), as well as changes in the incidence of pests and diseases (e.g. Porter et al., 1991). However, there is a growing body of evidence on the likely effects on grasslands and animal productivity (e.g. Baker et al., 1993; Parton et al., 1995; Rounsevell et al., 1996; Riedo et al., 1999; Luo et al., 2012; Schwerin, 2012). Effects on livestock primarily arise directly from heat stress (likely to be particularly significant in parts of the world where summer temperatures are already close to the maximum that livestock can tolerate) and indirectly via effects of climate change on grassland productivity and forage crop yields (IPCC, 2013). For cropping systems, changes in productivity are likely to result from direct effects of climate change at the plant level, or indirect effects at the system level, for example through shifts in nutrient cycling, and interactions between crops and weeds, pests and plant diseases (Fuhrer, 2003).

The key climatic factors most likely to pose threats to agriculture in combination with land degradation are: increases in the incidence and severity of droughts, heat stress, increased soil temperatures and changing evapotranspiration rates (Morton, 2007; D'Odorico et al., 2013; IPCC, 2013). These processes underpin a number of model-based predictions of the likely impacts of climate change on global agriculture. For example, putting aside the loss of low-lying land to sea level rise, Zhang and Cai (2011) estimated that the total global area of land suitable for agriculture is likely to contract by between 0.8–4.4 per cent by 2050, with the greatest contractions taking place in tropical and sub-tropical regions (up to 18 per cent reduction in Africa), due to changes in soil temperature and humidity. Jones and Thornton (2003) suggested there may be a 10 per cent drop in global maize production by 2055 due to climate change, with the greatest impacts likely to be felt in smallholder rainfed farms in Africa and Latin America (D'Odorico et al., 2013). Larger losses have been anticipated by other authors (e.g. see Hanjra and Qureshi, 2010, for a review). Similar concerns have been expressed for global wheat yields (Ortiz et al., 2008). Of course, accurate prediction is impossible in such complex systems, but these studies suggest that the effects of climate change on agricultural production are likely to be greatest in mid-latitude countries, which include some of the driest parts of the world at greatest risk from land degradation, and which host some of the poorest and most rapidly growing populations (Lee, 2009). In Europe, countries in the south that are already experiencing land degradation are likely to experience most disadvantages from climate change, with increases in water shortages and extreme weather events leading to lower, more variable yields and land abandonment in some areas (Olesen and Bindi, 2002).

When considering the likely interactions between climate change and land degradation, it is important to try and understand the balance between factors that may both increase and reduce the productive potential of the land. $CO_2$ and temperature are key variables affecting plant growth, development and function. Elevated $CO_2$ will directly influence plant physiology through its effect on photosynthesis, transpiration and respiration. However, rising temperature will have contrasting influences on these primary processes. There is now extensive literature on the direct effects of elevated $CO_2$ at levels ranging from the molecular to the global, which has substantially increased our knowledge of plant responses to rising $CO_2$ per se. It is well known that elevated $CO_2$ increases plant growth by an average of 20–40 per cent, and can compensate for environmental stress-induced reduction in growth by improving whole plant wateruse efficiency (Centritto et al., 1999a, 1999b, 2002). For example, it has been claimed that the $CO_2$ fertilization effect can increase photosynthesis and plant development rates (Kimball, 1983), and may help reduce drought stress by enabling plants to use water more efficiently (by reducing stomatal conductance) (Ghannoum et al., 2000; Leakey et al., 2006). Some research has suggested that many crops may be able to retrieve nutrients more effectively from the soil under elevated $CO_2$ (Drake et al., 1997), and yet there is also evidence that many important staple crops are likely to be less nutritious under elevated $CO_2$, with decreases in the zinc and iron content of wheat, rice, field peas and soybeans at $CO_2$ concentrations of 550 ppm, which are expected by 2050 (Myers et al., 2014). Elevated $CO_2$ is more likely to have positive effects on the yields of C3 crops, such as rice, wheat and soybean, through better use of resources and improved competition with C4 weeds, such as Cyperus rotundus (coco-grass) in rice (14 of the world's 17 most damaging terrestrial weed species are C4 plants in C3 crops according to Morison (1989)). Of course, vice versa, C3 weeds such as Striga spp. are likely to have an increased competitive advantage over C4 crops such as sorghum and pearl millet, which are important staples in drier parts of sub-Saharan Africa (Box 4.1). Positive effects on crops may also come from reduced susceptibility to the negative effects of ozone and improved pest and disease resistance (Fuhrer, 2003).

However, many of these beneficial effects may be offset by the negative effects of a warmer climate and changes in precipitation. The effects of warming on ecosystems will be more complex, in time and space, than the response to elevated

---

## BOX 4.1: *STRIGA*

*Striga* weeds belong to the family of *Scrophulariaceae*. They grow by parasitising some of the world's most important food crops in order to supply themselves with water and nutrients (Press and Gurney, 2000). *Striga* spp. can be found across the world and threaten 44 million hectares of arable land, as well as the livelihoods of more than 100 million farmers. In Africa, at least 35 different species of this weed are found. *Striga* reduces the productivity of a range of grain crops including maize, sorghum, millet and dryland rice by up to 50 per cent, to the extent that Nickrent and Musselman (2004) consider this genus of parasite has a bigger impact on humans and agroecosystems than any other parasitic plant species.

Part of the danger of *S. asiatica* is that it grows well in areas receiving less than 1500 mm of rain per annum, and prefers soils with low organic matter contents, low nitrogen levels and significant stone and gravel fractions (Stringer *et al.*, 2007). Climate changes, combined with land degradation, are likely to suit the proliferation of *S. asiatica* in a number of locations into the future. This is problematic, not least because *S. asiatica* is so difficult to eradicate. Farmers are often unaware of its presence, because after germination, the parasite develops under the ground for 4–6 weeks, during which time it relies on its host plant for water and nutrients. By the time the weed is visible at the soil surface, the damage to the host plant has already taken place, so weeding is not particularly effective as a control measure. Also, a single *S. asiatica* plant can also produce more than 100,000 tiny seeds. These are easily dispersed by wind and can remain dormant in the soil for up to 20 years before attacking the crop. This means that even if weeding does take place, large dormant seed banks remain. Soil fertility management is considered an important way to manage *S. asiatica*, as high levels of nitrogen in the form of ammonium sulphate and urea are toxic to the weeds, although small increases in nitrogen in very low fertility soils can actually benefit the parasite. While *Striga* spp. has important impacts on today's rural livelihoods and agricultural production, there is considerable potential for the problem to worsen in the future if it is not adequately tackled.

$CO_2$ concentration, because temperature impacts virtually all chemical and bio-logical processes, whereas the direct influence of $CO_2$ is almost entirely limited to leaves (Centritto *et al.*, 2011b). Benefits of elevated $CO_2$ for C3 crops are likely to be offset to an extent by increased competitive advantage of C4 weeds and insect damage in a warmer climate. Similarly in grasslands, elevated $CO_2$ increases dry matter production (especially for nitrogen-fixing legumes), but these benefits are likely to be offset by increased temperatures, for example due to increased insect damage. Although elevated $CO_2$ concentrations in the atmosphere have a fertilizing effect on crops, elevated tropospheric ozone levels can have a damaging effect on crop yields, and there is an uncertain relationship between $CO_2$ and ozone, mean temperature, extremes, water, nitrogen and land degradation processes (Heagle, 1989; IPCC, 2013). Increases in rainfall intensity are likely to increase rates of water erosion, which will be exacerbated by low vegetation cover as a result of land degradation (IPCC, 2013). Ultimately, the effects of climate change are more likely to be dominated by changes in temperature and precipitation than they are on elevated $CO_2$: "agroecosystem responses will be dominated by [impacts] caused directly or indirectly by shifts in climate, associated with altered weather patterns, and not by elevated $CO_2$ per se" (Fuhrer, 2003: 1).

With this in mind, it is likely that the effects of climate change on agricultural yields will vary geographically, depending on the factors that currently limit crop yields and how close temperatures are to critical thresholds for plant or animal growth. For example, where temperature is currently a limiting factor in plant productivity, increases in temperature may provide more optimal growing condi-tions for some crops and lengthen growing seasons, leading to increased productiv-ity. However, where crops are already growing at optimal temperatures, increased temperatures may lead to heat stress and reduce productivity. For example, within a range of 10–35°C, increases in ambient temperature enable maize crops to develop more quickly, and so complete phenological stages in shorter periods of time, but reductions in the rate of development have been noticed between 35–41°C (Yan and Hunt, 1999). Even within the same climatic zone, the effects of changing precipitation will be mediated by soil type, with crops growing on soils with higher water holding capacity able to buffer the effects of a lower and/or more sporadic rainfall more effectively than freely draining, sandy soils with limited water holding capacity. The cropping system and other land management adaptations can also strongly mediate the effects of climate change, further spatially differentiating impacts. Although it is difficult to predict impacts precisely due to these mediating factors, it is likely that areas where crops are already experiencing water stress will face an increased likelihood of crop failure under future climatic conditions (Challinor *et al.*, 2007).

## 4.3 Sensitivity of forests

Forests are particularly vulnerable to climate change, because their long life-span does not enable them to adapt rapidly to environmental changes (Lindner *et al.*,

2010). In addition to the provision of timber, forests provide many of the poorest populations in the world with non-timber forest products, which form an important component of their livelihoods (Gitay et al., 2001; Shvidenko et al., 2005). The degradation and destruction of forests is an important cause of land degradation around the world, increasing the sensitivity of soils to erosion, which can potentially lead to the long-term loss of productivity (MA, 2005a). The principal mechanisms through which climate change is likely to affect forests are the effects of elevated $CO_2$ and ozone, the effects of increased temperatures and altered precipitation regimes on tree growth and susceptibility to wildfire and disease.

Forest ecosystems will experience a combination of numerous environmental stresses, which may significantly alter their physiological feedback on climate, through evapotranspiration, albedo and carbon cycling. The natural biogeochemical movement of carbon to and from the terrestrial vegetation is larger than that from anthropogenic activities. Forests also play a major role in regulating the global hydrological cycle. Together with carbon sequestration, evapotranspiration, through feedbacks with clouds and precipitation, exerts a negative "physiological" forcing on both the regional and continental climate. Climate change may critically alter the biogeophysical and biogeochemical functioning of forests (Bonan, 2008; Rotenberg and Yakir, 2010). However, the current ability to predict when regional-scale plant stress will exceed a threshold that results in rapid and large-scale shifts in ecosystem structure and function is lacking. Thus, it is fundamental to assess potential climate change impacts, including changes in vegetation and associated ecosystem services, in order to predict the future feedbacks to the climate system.

The extent to which elevated $CO_2$ is likely to increase the growth of trees will depend on other limiting factors such as nutrient availability (Saxe et al., 1998; Norby et al., 1999; Ainsworth and Long, 2005), with nitrogen-fixing trees more likely to respond to elevated $CO_2$ than other species of tree (Hungate et al., 2003; Luo et al., 2004). Increased allocation of carbon to root growth may enable plants to exploit deeper soil water and ameliorate some of the effects of reduced water availability under climate change (Wullschleger et al., 2002). However, increased concentrations of ground-level ozone, resulting from chemical reactions between Nitrogen Oxides ($NO^x$) and Volatile Organic Compounds (VOCs) in the presence of sunlight, are likely to increase drought stress in trees (McLaughlin et al., 2007). Emissions from industry, vehicle exhausts, chemical solvents and energy production are some of the major $NO^x$ and VOC sources.

Although effects would vary between locations depending on site-specific factors, in general increases in temperature would likely benefit trees at higher latitudes, extending growing seasons and increasing rates of photosynthesis (Lindner et al., 2010). However, in drier climates where water already limits tree growth, heat stress is likely to inhibit photosynthesis, leading to stunted growth. An increased incidence and severity of droughts is likely to increase the likelihood, incidence and severity of wildfires (Lindner et al., 2010; for example, see Canadian studies by Podur et al., 2002; Gillett et al., 2004). The likely impacts of drought on forests has been projected for several regions (e.g. the Amazon and Europe; Cox *et al.*, 2004; Schaphoff *et al.*, 2006; Scholze *et al.*, 2006), showing impacts on forest net ecosystem productivity and wildfire risk, with the potential for a positive feedback to climate change through the release of carbon to the atmosphere and influences on regional climate. Dryland forests are less vulnerable to an increase in the incidence and severity of wildfires than forests in other areas, because they are usually already well adapted to cope with fire (Gonzalez *et al.*, 2010, see Box 4.2). However, any reduction in forest cover has the potential to contribute towards land degradation, unless it is rapidly replaced with alternative vegetation cover to prevent physical and chemical soil degradation and maintain the productive potential of the land.

An increase in the intensity of storms may lead to increased wind-throw of trees, reducing the amount of timber that can be recovered from forests (Lindner *et al.*, 2010), and increased flooding may adversely affect riparian forests (Glenz *et al.*, 2006;

## BOX 4.2: DRYLAND TREES AND THEIR ADAPTATIONS TO FIRE

Fire plays a vital role within many ecosystems. Indeed, grasslands and savannas that dominate much of the world's land surface would not even exist without the presence of wildfire. Similarly, many dryland tree species have adapted in order to cope with wildfires. While the specific impact of fire on plant evolution is difficult to trace, researchers have been able to identify particular fire-adaptive genetic traits in some pine species (*Pinus* spp.). These include bark thickness, serotiny (the release of seeds in response to an environmental stimulus), branch shedding, presence of a grass stage and re-sprouting capacity (He *et al.*, 2012). Taking each of these in turn, thicker bark insulates against fire, protecting the tree from the heat. This is a particularly useful trait for species in areas with ground level fires. Some types of pine trees retain mature cones that open following exposure to high temperatures associated with fire. This allows rapid and prolific seed release at a time when competition for the resources needed for successful germination (light, water, nutrients) is lowest. Other species shed their lower branches after their foliage has senesced, allowing a gap to present between the ground and the crown of the tree. A gap is useful as it reduces the chances of surface fires spreading into the tree canopy. "Grass stage" refers to a delay in trunk development whereby the bud of the pine tree is protected by a dense constellation of needles at ground level. During fires, the needles are burnt but regrow, while the growing tip is protected. Finally, some dryland tree species have evolved to re-sprout from their underground organs following exposure to fire. While these adaptations have proved successful in managing the effects of fire regimes to date, less is known about their abilities to cope with a changed regime resulting from climate changes, which could involve more intense and more frequent fires.

Kramer *et al.*, 2008). In forests where frosts and generally low temperatures currently limit insect outbreaks (Virtanen *et al.*, 1996; Volney and Fleming, 2000), climate warming may lead to more insect problems in future (Carroll *et al.*, 2004). If trees are already drought-stressed, these insect outbreaks are more likely to lead to tree mortality (Logan *et al.*, 2003; Gan, 2004). It is vital that future climate impacts are considered not just on existing forests, but also when deciding where to plant new trees (Box 4.3).

---

## BOX 4.3: FORESTS, TREE PLANTING AND CLIMATE ANALOGUES

Tree planting, afforestation and reforestation efforts have received considerable attention as a route towards the restoration of degraded forest lands, the prevention of land degradation and as a way of helping to mitigate climate change. However, the impacts of climate change on particular tree species are not always taken into account when deciding which trees to plant, and where to plant them. Partners working as part of the CGIAR's Research Programme on Climate Change, Agriculture and Food Security (CCAFS) assert that strategies and technologies for adapting to climate change in particular locations should ideally be grounded in knowledge of the future climatic conditions in those locations (Ramirez-Villegas *et al.*, 2011). As a consequence, they have developed a climate analogues tool (www.ccafs-analogues.org/tool/). The tool considers climate dissimilarities in order to identify which parts of the world have a climate that is similar to the current or future climate of a reference site. It also can include other variables such as soils, crops and socio-economic indicators. When it comes to tree planting, the tool can help to select areas that are most likely to be suitable for particular species, taking into account factors like their volume, resistance to disease etc., whilst also identifying which kinds of trees are most likely to be resilient to climate change in ten years' time.

The tool is also useful for identifying agricultural adaptation options. It can be used to identify historical adaptation trajectories and to help capture the real world capacity of farmers to adapt. The CGIAR argues that by linking up present-day farming systems with their possible future analogues, it can help to enable knowledge exchange between farmers in locations that share common climate interests by promoting learning about adaptation options. Through the 'Farms of the Future' project farmers can participate in exchange visits to experience the climate challenges they are projected to face in the future. The analogues tool also permits the testing and validation of adaptation strategies and technologies. For example, it can help identify the types of crop varieties and traits that will be needed to cope with climate and land degradation induced stresses, such as changing salinity levels. Also, by pinpointing where crop diversity will be at most risk, it allows prioritization of the collection and conservation of genetic resources.

Although forest ecosystem responses to elevated $CO_2$ have been well studied, results from forest experiments do not necessarily translate to the world's drylands. For example, the relationship between $CO_2$ enhancement of aboveground productivity and precipitation is fundamentally different for forests and for drylands (shrublands and grasslands) (Nowak *et al.*, 2004). Drylands are pulsed systems with high temporal and spatial variability in availability of multiple resources, especially water. Thus, increased photosynthesis and water use efficiency with increased atmospheric $CO_2$ do not necessarily increase dryland biomass production (Newingham *et al.*, 2013) even though carbon sequestration occurs (Evans *et al.*, 2014).

## 4.4 Sensitivity of freshwater provision

Impacts of climate change on freshwater resources are likely to increase significantly with increasing GHG concentrations, according to the IPCC (2014). Existing water shortages are increasing in response to the combined effects of climate change, land degradation, land cover change and population increase (MA, 2005a). The MA (2005a) stated with "a high degree of certainty" that these pressures would lead to "an accelerated decline in water availability and biological production in drylands". From 1960–2000, global use of fresh water including regions affected by DLDD increased at a mean rate of 25 per cent per decade (MA, 2005a). Water availability in drylands is projected to decrease further from an average of 1,300 $m^3$ per person per year in 2000, which was already below the 2,000 cubic meters required for minimum human well-being according to the MA (2005a). Modelling studies have suggested that for each degree of climate warming, approximately 7 per cent of the global population will be exposed to a decrease of renewable water resources of at least 20 per cent.

IPCC (2013) presents robust evidence that climate change will significantly reduce renewable surface water and groundwater resources in most dry subtropical regions. It is also likely to reduce water quality due to increases in sediment, nutrient and pollutant loadings as a result of increased rainfall intensity and reduced dilution of pollutants during droughts (IPCC, 2013). This is likely to have implications for agriculture, particularly livestock production with further significant effects on energy and food security.

Effects of climate change on renewable surface water and groundwater resources are likely to be significant in most dry subtropical regions, intensifying competition for water among agriculture, ecosystems, settlements, industry and energy production. In addition to reducing streamflow, climate change is anticipated to significantly compromise water quality in many areas, posing risks to drinking water quality (even in the presence of water treatment). These risks arise primarily from increases in sediment, nutrient and pollutant loadings (due to heavy rainfall), increased concentration of pollutants during droughts, and disruption of water treatment facilities during floods. Increases in heavy rainfall and temperature are projected to increase soil erosion and sediment yields, although the extent of these changes will depend on the seasonality of rainfall, land cover and soil management practices.

Unless GHG emissions are reduced significantly, by the end of the twenty-first century there will be three times more people exposed every year to severe flooding (of the magnitude typically only experienced every 100 years during the twentieth century). Effects of climate change on freshwater resources are already being felt in different ways around the world. For example, in Africa, the Zambezi River flooded in Mozambique in 2008, displacing 90,000 people, with approximately one million people living in the flood-affected areas. In 2011, floods in Australia and New Zealand caused severe damage to infrastructure and settlements, as well as 35 deaths. Lack of water is problematic too. A heat wave in Victoria, Australia, in 2009 was associated with more than 300 deaths, while intense bushfires destroyed more than 2000 buildings and led to 173 deaths. Droughts in southeast Australia (1997–2009) and many parts of New Zealand (2007–2009; 2012–2013) have led to significant economic losses. Regional GDP in the southern Murray-Darling Basin was below levels that were forecast by about 5.7 per cent in 2007–2008, while the agriculture sector in New Zealand lost about NZ$3.6 between 2007 and 2009.

## 4.5 Sensitivity of livelihoods

Interactions between climate change and land degradation have the potential to significantly affect livelihoods through their effects on provisioning services from agricultural, forestry and fresh water systems. The key climatic processes that are likely to interact with land degradation processes to threaten livelihoods are droughts, heat stress, and increased soil temperatures and evapotranspiration rates. These interactions are likely to particularly constrain the livelihoods of those most dependent on agriculture and natural resources, especially in contexts where adaptive capacity is low. Adaptive capacity may, for example, be limited by a lack of physical, human or financial assets in disadvantaged areas, or further constrained by a lack of effective governance[1] or a lack of incentives. Governance structures are needed to provide access to information and build capacity for coordinated action to adapt to the effects of land degradation and climate change. Furthermore, policy instruments including regulatory mechanisms (e.g. prohibition or zoning of particular land uses), financial mechanisms (e.g. incentives or taxes) and the creation of new markets (e.g. payments for ecosystem services) may in some contexts be able to reduce sensitivity and increase the adaptive capacity of ecosystems and populations to climate change and land degradation.

Although many of the effects of climate change and land degradation on livelihoods will take place via changes in the provision of ecosystem services, both processes can directly affect livelihoods and human well-being. This may occur indirectly through the effects of climate change and land degradation on natural capital and subsequent effects on physical, human, social and financial capital, and can include important negative effects on economies. However, climate change and land degradation may also directly affect these other assets (e.g. weakening social networks through heat and disease-vector related illness and mortality), compromising financial assets (e.g. due to reduced agricultural productivity or failed harvests) or rendering physical infrastructure (e.g. existing flood defenses) obsolete (Reed et al., 2013). The impacts of climate change and land degradation on livelihoods can further impact upon adaptation in ways that can have disproportionate effects upon women (Box 4.4). It is also unclear whether water shortages under climate change are more likely to drive conflict or cooperation (Box 4.5).

---

## BOX 4.4: LIVELIHOODS AND ADAPTATION

One of the key routes towards livelihood adaptation is for people to diversify their livelihood activities. In doing so, they spread the risks they face over a wider range of options. Livelihood diversification can take many forms, from broadening the range or varieties of crops grown within an agricultural diversification strategy (often involving shifts to cash crops such as cotton, sugar and tobacco) to undertaking entirely new livelihood activities. Such new activities might include: temporary migration to engage in paid agricultural work with associated remittances being returned home; engagement in paid work in sectors other than agriculture, such as in mining or forestry; tourism sector activities; handicrafts; fishing; food processing activities (for example, making and selling shea butter or millet beer) and hawking of both goods and services (e.g. shoe shining, windscreen washing). While livelihood diversification can spread risks, some authors have argued that it can increase risks, particularly for women. Bryceson and Fonseca (2006) report on *ganyu* as an important form of livelihood diversification in Malawi, particularly during times of crisis. *Ganyu* is the exchange of labour for goods, services or cash and has a long history in Malawi. It is particularly common during busy times of the agricultural calendar, and causes poorer families to divert their labour away from their own fields at a time when they can least afford it. At the same time, reliance on *ganyu* has been accompanied with increasingly unfavourable terms of exchange. Many agreements are made by word of mouth and are insecure, with the terms and conditions being defined by the more powerful Parties. For example, researchers note that sexual activity has been progressively incorporated into women's *ganyu* contracts, and that this is a particular issue for those in most need of money and food. This is not the same as "prostitution" which involves active solicitation of sex as part of a service market (Campbell, 2003). *Ganyu* transactional sex is more down to chance encounters and driven by a household's need for food, whereby women are poorly equipped to set out the terms. In the context of the HIV/AIDS pandemic in sub-Saharan Africa, such livelihood diversification and risk spreading through *ganyu* can leave vulnerable groups open to other risks that have far-reaching implications. Adaptations to climate change and land degradation that increase vulnerability in this way can be viewed as maladaptations.

## BOX 4.5: WATER WARS?

The jury is still out on whether land and water degradation combined with climate change is likely to lead to conflict or cooperation. In a study by Yoffe *et al.* (2003), instances of cooperation over shared water resources far outnumbered incidences of conflict. Similarly, Postel and Wolf (2001) noted that treaties signed by riparian countries often help to reduce competing claims and to promote cooperation over water resource use and allocation. These two studies present water as a "connector" rather than as a "divider" and as a catalyst for unity and interdependence across many scales.

A contrary view argues that given the increasing water scarcity, particularly under climate change, and because there is no suitable substitute for most of our uses of water, a link exists between water scarcity and conflict (e.g. Bernauer and Siegfried, 2012; Hauge and Ellingsen, 1998; Kreamer, 2012). Lowi (1999) used case studies from the Middle East and South Asia to show that competition over scarce fresh water leads to severe socio-political violence. Butts (1997) found that the Earth's history is replete with instances of violent water conflicts, while Eriksson *et al.* (2003) noted that from 1989–2003, water scarcity was linked to 80 to 90 per cent of recorded internal armed conflicts globally. This is consistent with Gleick and Heberger's (2012) findings that Israel, Jordan and Palestine – all dryland countries with concerns about water – fired gunshots, burned houses, blew up dams and undertook some form of water-related military and political demonstration in the second half of the twentieth century. In Somalia over 250 lives were lost during a series of water clashes in Rabdore village, during the region's relentless 3-year drought from 2002–2005 (Wax and Thompson, 2006).

From the literature, three key linkages between water and conflict are noticeable. First, water conflict can materialize when tensions involve water access and allocation. Allocating water among different users and uses, even when water is abundant, can be a cause of violence. Similarly, a decline in water quality, which may threaten human health, can be a source of potential aggression. Decline in volume or quality can induce mass migration which can economically, socially and politically destabilize destination locations (Carius *et al.*, 2004). Second, the importance of water in sustaining human well-being, mostly through agricultural livelihoods, provides a pathway to conflict (Conca, 2006). Livelihood loss can lead to deprivation, which is a well-known traditional driver of conflict and violence. Third, there are examples of places where power struggles and inadequate water governance affect the potential for conflict. This is particularly the case in settings where rivers cross national boundaries and where water institutions lack human, technical and administrative capacities (Ludwig *et al.*, 2011).

Based on Okpara *et al.* (2015).

## 4.6 Synthesis

This chapter has considered the likely exposure and sensitivity of provisioning ecosystem services to climate change and land degradation processes, and the consequences that this may have for food security and the sustainability of livelihoods dependent on these services in regions affected by DLDD. Exposure to the combined processes of climate change and land degradation is likely to be highly variable across different regions and systems, and there are a number of important uncertainties, for example linked to the range of different and sometimes opposing feedbacks between the two processes. This makes it difficult to accurately determine the likely sensitivity of human populations and ecosystems to climate change and land degradation.

The remainder of the chapter has outlined what we currently know about the sensitivity of provisioning services from agriculture and forests, and fresh water, to climate change and land degradation. Impacts on agricultural systems are likely to be diverse, including some improvements to provisioning services, mainly related to the $CO_2$ fertilization effect and certain aspects of climate change. However, when considering all likely climatic effects, both direct (e.g. related to drought and heat) and indirect (e.g. linked to the spread of diseases and pests), and likely interactions with land degradation, overall effects on provisioning services are likely to be negative, with knock-on effects for food security and the sustainability of livelihoods in many regions affected by DLDD. The picture is similar for forests, with dryland forests and the communities that live near them at particular risk from an increased risk of wildfire, which typically then leads to a number of irreversible land degradation processes. Water shortages are increasing in response to the combined effects of climate change, land degradation and population increases.

Interactions between climate change and degradation may threaten livelihoods as they compromise the supply of provisioning services from agriculture, forests and fresh water systems. The key climatic processes that are likely to interact with land degradation processes to threaten livelihoods are droughts, heat stress, and increased soil temperatures and evapotranspiration rates. These interactions are likely to particularly constrain the livelihoods of those most dependent on agriculture and natural resources, especially in contexts where adaptive capacity is low.

## Note

1 MA (2005b: 77) define governance as, "the sum of the many ways in which individuals and institutions, public and private, manage issues".

## References

Ainsworth, E.A., Long, S.P. 2005. What have we learned from 15 years of free-air $CO_2$ enrichment (FACE)? A meta-analytic review of the responses of photosynthesis, canopy properties and plant production to rising $CO_2$. *New Phytologist* 165: 351–372.

Baker, B.B., Hanson, J.D., Bourdon, R.M., Eckert, J.B. 1993. The potential effects of climate change on ecosystem processes and cattle production on U.S. rangelands. *Climatic Change* 25: 97–117.

Bernauer, T., Siegfried, T. 2012. Climate change and international water conflict in Central Asia. *Journal of Peace Research* 49(1): 227–239.

Bonan, G.B. 2008. Forests and climate change: forcings, feedbacks, and the climate benefits of forests. *Science* 320: 1444–1449.

Bryceson, D.F., Fonseca, J. 2006. An enduring or dying peasantry? Interactive impact of famine and HIV/AIDS in rural Malawi. In: *AIDS, Poverty and Hunger: Challenges and Responses*, Gillespie, S. (ed.). International Food Policy Research Institute, Washington D.C., USA, pp. 97–108.

Butts, K. 1997. The strategic importance of water. *Parameters*, Spring 1997, pp. 65–83.

Campbell, C. 2003. *Letting Them Die: Why HIV/AIDS prevention programmes fail*. Oxford: James Currey.

Carius, A., Dabelko, D., Wolf, A. 2004. Water, conflict, and cooperation. *ECSP Report* (10). Wilson Center, USA.

Carroll, A.L., Taylor, S.W., Régnière, J., Safranyik, L. 2004. Effects of climate change on range expansion by the mountain pine beetle in British Columbia. In: *Mountain Pine Beetle Symposium: Challenges and solutions*, Shore, T.L., Brooks, J.E., Stone, J.E. (eds). *Natural Resources Canada, Canadian Forest Service, Pacific Forestry Centre*. Victoria: British Columbia, pp. 223–232.

Centritto, M., Lee, H.S.J., Jarvis, P.G. 1999a. Interactive effects of elevated $CO_2$ and drought on cherry (*Prunus avium*) seedlings. In: Growth, whole-plant water use efficiency and water loss. *New Phytologist* 141: 129–140.

Centritto, M., Lee, H. S. J., Jarvis, P. G. 1999b. Increased growth in elevated $CO_2$: an early, short-term response? *Global Change Biology* 5: 623–633.

Centritto, M., Lucas, M.E., Jarvis, P.G. 2002. Gas exchange, biomass, whole-plant water-use efficiency and water uptake of peach (*Prunus persica*) seedlings in response to elevated carbon dioxide concentration and water availability. *Tree Physiology* 22: 699–706.

Centritto, M., Tognetti, R., Leitgeb, E., Strřelcová, K., Cohen, S. 2011a. Above ground processes – anticipating climate change influences. In: Forest management and the water cycle: an ecosystem-based approach, Bredemeier, M., Cohen, S., Godbold, D.L., Lode, E., Pichler, V., Schleppi, P. (eds). *Ecological Studies* 212: 31–64.

Centritto, M., Brilli, F., Fodale, R. Loreto, F. 2011b. Different sensitivity of isoprene emission, respiration, and photosynthesis to high growth temperature coupled with drought stress in black poplar (*Populus nigra*). *Tree Physiology* 31: 275–286.

Challinor, A.J., Wheeler, T.R., Craufurd, P.Q., Ferro, C.A.T., Stephenson, D.B. 2007. Adaptation of crops to climate change through genotypic responses to mean and extreme temperatures. *Agriculture, Ecosystems & Environment* 119: 190–204.

Conca, B.K. 2006. The new face of water conflict. *Navigating Peace*, Issue 3, Wilson Center, USA.

Cox, P.M., Betts, R.A., Collins, M., Harris, P.P., Huntingford, C., Jones, C.D. 2004 Amazonian forest dieback under climate-carbon cycle projections for the 21st century. *Theoretical Applications of Climatology* 78: 137–156.

Dai, A. 2011. Drought under global warming: a review. *Wiley Interdisciplinary Reviews: Climate Change* 2: 45–65.

D'Odorico, P., Bhattachan, A., Davis, K.F., Ravi, S., Runyan, C.W. 2013. Global desertification: drivers and feedbacks. *Advances in Water Resources* 51: 326–344.

Drake, B.J., Gonzàlez-Meler, M.A., Long, S.P. 1997. More efficient plants: a consequence of rising atmospheric $CO_2$? *Annual Review of Plant Physiology and Plant Molecular Biology* 48: 609–639.

Eriksson, M., Wallensteen, P., Sollenberg, M. 2003. Armed Conflict, 1989–2002. *Journal of Peace Research* 40(5): 593–607.

Evans, R.D., Koyama, A., Sonderegger, D.L., Charlet, T.N., Newingham, B.A., Fenstermaker, L.F., Harlow, B., Jin, V.L., Ogle, K., Smith, S.D., Nowak, R.S. 2014. Greater ecosystem carbon in the Mojave Desert after ten years exposure to elevated $CO_2$. *Nature Climate Change* 4: 394–397.

Fischer, G., Shah, M., van Velthuizen, H., Nachtergaele, F. 2006. Agro-ecological zones assessment. In: *Land Use, Land Cover and Soil Sciences/EOLSS*, W.H. Verheye (ed.). Eolss Publishers: Oxford, UK.

Fuhrer, J. 2003. Agroecosystem responses to combination of elevated $CO_2$, ozone, and global climate change. *Agriculture, Ecosystems & Environment* 97: 1–20.

Gan, J.B. 2004. Risk and damage of southern pine beetle outbreaks under global climate change. *Forest Ecology and Management* 191: 61–71.

Ghannoum, O., von Caemmerer, S., Ziska, L.H., Conroy, J.P. 2000. The growth response of C4 plants to rising atmospheric $CO_2$ partial pressure: a reassessment. *Plant Cell & Environment* 23: 931–942.

Gillett, N.P., Weaver, A.J., Zwiers, F.W., Flannigan, M.D. 2004. Detecting the effect of climate change on Canadian forest fires. *Geophysical Research Letters* 31: L18211.

Gitay, H., Brown, S., Easterling, W., Jallow, B. 2001. Ecosystems and their goods and services. In: *Climate change 2001: impacts, adaptation, and vulnerability. Contribution of Working Group II to the third assessment report of the intergovernmental panel on climate change*, McCarthy, J.J., Canziani, O.F., Leary, N.A., Dokken, D.J., White, K.S. (eds). Cambridge University Press: Cambridge, 237–342.

Gleick, P.H., Heberger, M. 2012. Water conflict chronology. *The Worlds Water*, 7: 175–214.

Glenz, C., Schlaepfer, R.I.I., Kienast. F. 2006. Flooding tolerance of Central European tree and shrub species. *Forest Ecology and Management* 235: 1–13.

Gonzalez, P., Neilson, R.P., Lenihan, J.M., Drapek, R.J. 2010. Global patterns in the vulnerability of ecosystems to vegetation shifts due to climate change. *Global Ecology and Biogeography* 19: 755–768.

Hanjra, M.A., Qureshi, M.E. 2010. Global water crisis and future food security in an era of climate change. *Food Policy* 35: 365–377.

Hauge, W, Ellingsen, T. 1998. Beyond environmental scarcity: causal pathways to conflict. *Journal of Peace Research* 35(3): 299–317.

He, T., Pausas, J.G., Belcher, C.M., Schwilk, D.W., Lamont, B.B. 2012. Fire adapted traits of Pinus arose in the fiery Cretaceous. *New Phytologist* 194: 751–759.

Heagle, A.S., Miller, J.E., Booker, F.L., Pursley, W.A. 1999. Ozone stress, carbon dioxide enrichment, and nitrogen fertility interactions in cotton. *Crop Science* 39: 731–741.

Hungate, B.A., Dukes, J.S., Shaw, M.R., Luo, Y., Field. C.B. 2003. Nitrogen and climate change. *Science* 302: 1512–1513.

IPCC. 2007. Climate Change 2007. The Physical Science Basis. In: *Contribution of Working Group I to the fourth assessment report of the intergovernmental panel on climate change*, Solomon, S., Qin, HD., Manning, M., Chen, Z., Marquis, M., Averyt, K.B., Tignor, M., Miller, H.L. (eds). Cambridge University Press: Cambridge, United Kingdom and New York, NY, USA, 996 pp.

IPCC. 2013. Climate Change 2013. The Physical Science Basis. In: *Contribution of Working Group I to the fifth assessment report of the intergovernmental panel on climate change*, Stocker, T.F., Qin, D., Plattner, G.K., Tignor, M., Allen, S.K., Boschung, J., Nauels, A., Xia, Y., Bex, V., Midgley, P.M. (eds). Cambridge University Press: Cambridge, United Kingdom and New York, NY, USA, 1535 pp.

IPCC. 2014. Climate Change 2014. Impacts, Adaptation, and Vulnerability. In: *Part A: global and sectoral aspects. Contribution of Working Group II to the fifth assessment report of the inter-governmental panel on climate change*, Field, C.B., Barros, V. R., Dokken, D. J., Mach, K. J., Mastrandrea, M. D., Bilir, T. E., Chatterjee, M., Ebi, K. L., Estrada, Y.O., Genova, R.C., Girma, B., Kissel, E.S., Levy, A.N., MacCracken, S., Mastrandrea, P. R., White, L. L. (eds). Cambridge University Press: Cambridge, United Kingdom and New York, NY, USA.

Jones, P.G., Thornton, P.K. 2003 The potential impacts of climate change on maize production in African and Latin America in 2055. *Global Environmental Change* 13: 51–59.

Kimball, B.A. 1983. Carbon dioxide and agricultural yield: an assemblage and analysis of 430 prior observations. *Agronomy Journal* 75: 779–788.

Kramer, K., Vreugdenhil, S.J., van der Werf, D.C. 2008. Effects of flooding on the recruitment, damage and mortality of riparian tree species: a field and simulation study on the Rhine floodplain. *Forest Ecology and Management* 255: 3893–3903.

Kreamer, D.K. 2012. The past, present, and future of water conflict and international security. *Journal of Contemporary Water Research & Education* 149(1): 87–95.

Leakey, A.D., Uribelarrea, M., Ainsworth, E., Naidu, S., Rogers, A., Ort, D., Long, S. 2006. Photosynthesis, productivity, and yield of maize are not affected by open-air elevation of $CO_2$ concentration in the absence of drought. *Plant Physiology* 140: 779–790.

Lee, H.L. 2009. The impact of climate change on global food supply and demand, food prices, and land use. *Paddy and Water Environment* 7: 321–331.

Lindner, M., Maroschek, M., Netherer, S., Kremer, A., Barbati, A., Garcia-Gonzalo, J., Seidl, R., Delzon, S., Corona, P., Kolstrom, M., Lexer, M.J., Marchetti, M. 2010. Climate change impacts, adaptive capacity and vulnerability of European forest ecosystems. *Forest Ecology and Management* 259: 698–709.

Logan, J.A., Regniere, J., Powell, J.A. 2003. Assessing the impacts of global warming on forest pest dynamics. *Frontiers in Ecology and Environment* 1: 130–137.

Lowi, M.R. 1999. Water and conflict in the Middle East and South Asia: are environmental issues and security issues linked? *The Journal of Environment & Development* 8(4): 376–396.

Ludwig, R., Roson, R., Zografos, C., Kallis, G. 2011. Towards an inter-disciplinary research agenda on climate change, water and security in Southern Europe and neighboring countries. *Environmental Science & Policy* 14(7): 794–803.

Luo, Y., Su, B., Currie, W.S., Dukes, J.S., Finzi, A., Hartwig, U., Hungate, B., McMurtrie, R.E., Oren, R., Parton, W.J., Pataki, D.E., Shaw, M.R., Zak, D.R., Field, C.B. 2004. Progressive nitrogen limitation of ecosystem responses to rising atmospheric carbon dioxide. *BioScience* 54: 731–739.

Luo L., Wang Z., Mao D., Lou Y., Ren C., Song K. 2012. Responses of grassland net primary productivity in western Songnen Plain of Northeast China to climate change and human activity. *Shengtaixue Zazhi* 31: 1533–1540.

MA (Millennium Ecosystem Assessment). 2005a. *Ecosystems and Human Well-being: Current state and trends assessment*. Island Press: Washington, D.C., USA.

MA. 2005b. *Ecosystems and Human Wellbeing: Policy response*. Island Press: Washington, D.C., USA.

McLaughlin, S.B., Wullschleger, S.D., Sun, G., Nosal, M. 2007. Interactive effects of ozone and climate on water use, soil moisture content and streamflow in a southern Appalachian forest in the USA. *New Phytologist* 174: 125–136.

Morison, J.I.L. 1989. Plant growth in increased atmospheric $CO_2$. In: *Carbon Dioxide and Other Greenhouse Gases: Climatic and associated impacts*; Fantechi, R., Ghazi, A. (eds). *CEC, Reidel*. Dordrecht, The Netherlands, pp. 228–244.

Morton J.F. 2007. The impact of climate change on smallholder and subsistence agriculture. *Proceedings of the National Academy of Sciences USA* 104: 19680–19685.

Myers, S.S., Zanobetti, A., Kloog, I., Huybers, P., Leakey, A.D.B., Bloom, A.J., Carlisle, E., Dietterich, L.H., Fitzgerald, G., Hasegawa, T., Holbrook, N.M., Nelson, R.L., Ottman, M.J., Raboy, V., Sakai, H., Sartor, K.A., Schwartz, J., Seneweera, S., Tausz, M., Usui, Y. 2014. Increasing $CO_2$ threatens human nutrition. *Nature* 510: 139–142.

Newingham, B.A., Vanier, C.H., Charlet, T.N., Ogle, K., Smith, S.D., Nowak, R.S. 2013. No cumulative effect of elevated $CO_2$ on perennial plant biomass after ten years in the Mojave Desert. *Global Change Biology* 19: 2168–2181.

Nickrent, D.L., Musselman, L.J. 2004. Introduction to parasitic flowering plants. *The Plant Health Instructor*. DOI: 10.1094/PHI-I-2004-0330-01.

Norby, R.J., Wullschleger, S.D., Gunderson, C.A., Johnson, D.W., Ceulemans, R. 1999. Tree responses to rising $CO_2$ in field experiments: implications for the future forest. *Plant, Cell and Environment* 22: 683–714.

Nowak, R.S., Ellsworth, D.S., Smith, S.D. 2004. Tansley review: functional responses of plants to elevated atmospheric $CO_2$ – do photosynthetic and productivity data from FACE experiments support early predictions? *New Phytologist* 162: 253–280.

Okpara, U.T., Stringer, L.C., Dougill, A.J., Bila, M.D. 2015. Conflicts about water in Lake Chad: are environmental, vulnerability and security issues linked? *Progress in Development Studies* 15 (4): 1–18.

Olesen, J.E., Bindi, M. 2002. Consequences of climate change for European agricultural productivity, land use and policy. *European Journal of Agronomy* 16: 239–262.

Ortiz, R., Sayre, K.D., Govaerts, B., Gupta, R., Subbarao, G.V., Ban, T., Reynolds, M. 2008. Climate change: can wheat beat the heat? *Agriculture, Ecosystems & Environment* 126: 46–58.

Parton, W.J., Scurlock, J.M.O., Ojima, D.S., Schimel, D.S., Hall, D.O. 1995. Impact of climate change on grassland production and soil carbon worldwide. *Global Change Biology* 1: 13–22.

Podur, J., Martell D.L., Knight, K. 2002. Statistical quality control analysis of forest fire activity in Canada. *Canadian Journal of Forest Research* 32: 195–205.

Porter, J.H., Parry, M.L., Carter, T.R. 1991. The potential effects of climatic change on agricultural insect pests. *Agricultural and Forest Meteorology* 57: 221–240.

Postel, S.L., Wolf, A.T. 2001. Dehydrating conflict. *Foreign Policy* 126: 60–67.

Press, M.C., Gurney, A.L. 2000. Plant eats plant: sap-feeding witchweeds and other parasitic angiosperms. *Biologist* 47: 189–193.

Ramírez-Villegas, J., Lau, C., Köhler, A.K., Signer, J., Jarvis, A., Arnell, N., Osborne, T., Hooker, J. 2011. Climate analogues: finding tomorrow's agriculture today. Working paper no. 12. *CGIAR Research Program on Climate Change, Agriculture and Food Security (CCAFS)*. Available online at: www.ccafs.cgiar.org

Reed, M.S., Podesta, G., Fazey, I., Beharry, N.C., Coen, R., Geeson, N., Hessel, R., Hubacek, K., Letson, D., Nainggolan, D., Prell, C., Psarra, D., Rickenbach, M.G., Schwilch, G., Stringer, L.C., Thomas, A.D. 2013. Combining analytical frameworks to assess livelihood vulnerability to climate change and analyse adaptation options. *Ecological Economics* 94: 66–77.

Reed, M.S, Stringer, L.C., Dougill, A.J., Perkins, J.S., Atlhopheng, J.R., Mulale, K., Favretto, N. 2015. Reorienting land degradation towards sustainable land management: linking sustainable livelihoods with ecosystem services in rangeland systems. *Journal of Environmental Management* 151: 472–485.

Reilly, J.M., Schimmelpfennig, D. 1999. Agricultural impact assessment, vulnerability, and the scope for adaptation. *Climatic Change* 43: 745–788.

Riedo, M., Gyalistras, D., Fischlin, A., Fuhrer, J. 1999. Using an ecosystem model linked to GCM-derived local weather scenarios to analyse effects of climate change and elevated $CO_2$ on dry matter production and partitioning, and water use in temperate managed grasslands. *Global Change Biology* 5: 213–223.

Rotenberg, E., Yakir, D. 2010. Contribution of semi-arid forests to the climate system. *Science* 327: 451–454.

Rounsevell, M.D.A., Brignall, A.P., Siddons, P.A. 1996. Potential climate change effects on the distribution of agricultural grasslands in England and Wales. *Soil Use and Management* 12: 44–51.

Saxe, H., Ellsworth, D.S., Heath, J. 1998. Tree and forest functioning in an enriched $CO_2$ atmosphere. *New Phytologist* 139: 395–436.

Schaphoff, S., Lucht, W., Gerten, D., Sitch, S., Cramer, W., Prentice, I.C. 2006. Terrestrial biosphere carbon storage under alternative climate projections. *Climatic Change* 74: 97–122.

Scholze, M., Knorr, W., Arnell, N.W., Prentice, I.C. 2006. A climate change risk analysis for world ecosystems. *Proceedings of the National Academy of Science USA* 103: 13116–13120.

Schwerin, M. 2012. Climate change as a challenge for future livestock farming in Germany and Central Europe. *Zuchtungskunde* 84: 103–128.

Shvidenko, A., Barber, C.V., Persson, R. 2005. Forest and woodland systems. In: *Ecosystems and Human Well-being: Volume 1: Current state and trends*, Hassan, R., Scholes, R., Ash, N. (eds). Island Press: Washington., D.C., USA, 585–621.

Stringer, L.C, Twyman, C., Thomas, D.S.G. 2007. Learning to reduce degradation on Swaziland's arable land: enhancing understandings of *Striga asiatica*. *Land Degradation and Development* 18, 163–177.

Virtanen, T., Neuvonen, S., Nikula, A., Varama, M., Niemelä, P. 1996. Climate change and the risks of Neodiprion sertifer outbreaks on Scots pine. *Silva Fennica* 30: 169–177.

Volney, W.J.A., Fleming, R.A. 2000. Climate change and impacts of boreal forest insects. *Agriculture, Ecosystems & Environment* 82: 283–294.

Wax, E., Thompson, R. 2006. Dying for water in Somalia's drought. *Washington Post,* April 4: 20–22.

Wiggs, G.F.S., Thomas, D.S.G., Bullard, J.E. 1995. Dune mobility and vegetation cover in the southwest Kalahari desert. *Earth Surface Processes and Landforms* 20: 515–529.

Wullschleger, S.D., Tschaplinski, T.J., Norby, R.J. 2002. Plant water relations at elevated $CO_2$ – implications for water-limited environments. *Plant, Cell and Environment* 25: 319–331.

Yan, W., Hunt, L.A. 1999. An equation for modelling the temperature response of plants using only the cardinal temperatures. *Annals of Botany* 84: 607–614.

Yoffe, S., Wolf, A.T., Giordano, M. 2003. Conflict and cooperation over international freshwater resources: indicators of basins at risk. *Journal of the American Water Resources Association*, 39(5): 1109–1126.

Zhang, X.A., Cai, X.M. 2011. Climate change impacts on global agricultural land availability. *Environmental Research Letters* 6: 014014.

# 5

# EXPOSURE AND SENSITIVITY OF OTHER ECOSYSTEM SERVICES AND FEEDBACKS BETWEEN CLIMATE CHANGE AND LAND DEGRADATION

## 5.1 Introduction

In addition to the impacts on provisioning services outlined in the preceding chapter, there are likely to be a number of impacts of climate change on the delivery of other ecosystem services in regions affected by DLDD, many of which may further affect the sustainability of natural resource-based livelihoods. If land degradation is conceptualized more broadly as a reduction in the capacity of the land to provide ecosystem services (Reed *et al.*, 2015), then these effects are integral to our understanding of the vulnerability of populations and ecosystems to the interactive effects of climate change and land degradation. Principal among these impacts are changes in:

- *Supporting services*: for example, effects on soil formation and conservation, nutrient cycling and primary production;
- *Regulating services*: for example, effects on water regulation, climate regulation and pollination;
- *Cultural services*: for example, effects on aesthetic, cultural and spiritual benefits from nature.

This chapter considers each of these types of other ecosystem service and their likely exposure and sensitivity to climate change and land degradation. The chapter then explores how interactions between climate change and land degradation may play out as feedbacks, for example between carbon and vegetation in soil, and between vegetation cover and biodiversity, with implications across each of the ecosystem services considered in this and the previous chapter.

## 5.2 Supporting ecosystem services

Supporting ecosystem services are likely to be exposed and sensitive to a number of interactions between climate change and land degradation in regions affected by DLDD. Climate change is expected to decrease rates of soil formation in a number of ways, for example via increased soil organic matter decomposition rates (IPCC, 2013). It is also expected that climate change will increase soil erosion rates, primarily through more frequent high intensity rainfall events with greater erosive power (Nearing, 2001; Pruski and Nearning, 2002). This is likely to interact with changes in temperature, solar radiation and atmospheric $CO_2$ concentrations, depending on how these influence plant biomass production and hence vegetation cover (Nearing et al., 2004). Reduced vegetation cover increases the likelihood of both water and wind erosion, and may in some cases lead to a loss of soil productivity, particularly in the absence of sustainable agronomic practices. Although physical and chemical erosion (nutrient and organic matter losses, salinization, pollution and acidification) are widely used as indicators of land degradation, the relationship between erosion and the productivity of land is complex. For example, there is evidence that the majority of wind erosion in the Kalahari results in the localized redistribution of soil and nutrients (for example collecting around the base of shrubs), with minimal loss of nutrients from the system (Schlesinger et al., 1990; Dougill and Thomas, 2002). Elsewhere, there is evidence of agricultural systems that have maintained yields despite significant levels of water erosion (e.g. Warren, 2002). Although the effects of elevated $CO_2$ on litter decomposition and soil fauna are said by IPCC's Working Group II to the Fourth Assessment Report (Fischlin, 2007: 226) to "seem species-specific and relatively minor", there are studies that suggest climate change may lead to changes in soil fungal communities that may have impacts on soil structure that, particularly if combined with unsustainable tillage and management practices, increase erosion risks (Zhang et al., 2005; Rillig et al., 2002; Fischlin et al., 2007).

Moreover, a 2.4°C warming leads to approximately a 20 per cent increase in soil respiration (Norby et al., 2007), although this is moderated to an extent by acclimatization of the soil microbial community to moderate increases in soil temperature (Luo et al., 2001). Although the temperature sensitivity of soil respiration is especially critical in semi-arid regions, little research has been carried out in these environments (for a meta-analysis, see Hamdi et al., 2013). Moreover, there are two different aspects: acclimatization to a shift of a few degrees, and acclimatization to extreme events. In the latter case, laboratory experiments using incubation of soils from Tunisia with temperatures up to 50°C showed that the main determinant of soil temperature sensitivity is the amount of labile carbon rather than microbial adaptation of soil respiration to temperature (Hamdi et al., 2011). The resulting loss of soil carbon to the atmosphere may have long-term implications for soil fertility, water holding capacity and crop growth, with consequences for rural livelihoods, as well as posing the risk of a positive feedback to climate warming (Neely et al., 2009).

Other aspects might be considered such as the behaviour of soil inorganic carbon under climate change, soil respiration and wet-dry cycles changes (Bernoux and Chevallier, 2014). Dryland soils contain large amounts of inorganic carbon in the form of carbonates. Almost 97 per cent of soil inorganic carbon stocks worldwide are in soils that receive less than 750 mm annual rainfall, and in these soils there can be between two and ten times as much inorganic carbon as there is soil organic carbon (Cerling, 1984). Despite their significance (global inorganic carbon reserves are estimated to be 950 Gt), there has been limited research into this type of carbon (Bernoux and Chevallier, 2014). Precipitation of $CO_2$ from the atmosphere into carbonates occurs via biological processes (primarily root micro-organisms) at a rate of 0.007–0.266 Gt carbon/year in arid and semi-arid regions (Lal and Bruce, 1999). There is at present contradictory evidence on the effects of land use and management on soil inorganic carbon storage, with some studies suggesting that inorganic carbon storage may be lost or enhanced via irrigation practices (Bernoux and Chevallier, 2014). Wu et al. (2009) showed that in China, soil inorganic carbon loss occurred across all types of agricultural land uses, but was greatest in non-irrigated fields, compared to irrigated or paddy fields. Likely effects of climate change are uncertain, as soil inorganic carbon levels are positively correlated with temperature and negatively correlated with precipitation (Schlesinger, 1982; Wu et al., 2009).

More broadly however, soil organic matter stores are known to decline under elevated temperatures, due to increased respiration from the soil microbial community, converting organic carbon stored in the soil to $CO_2$ (Bond-Lamberty et al., 2004; Bond-Lamberty and Thomson, 2010; Hamdi et al., 2013). The rate of soil organic carbon sequestration may be enhanced through soil and water conservation strategies e.g. soil restoration and woodland regeneration, no-till farming, cover crops, nutrient management, manuring and sludge application, improved grazing, water conservation and harvesting, efficient irrigation, agroforestry practices, and growing energy crops on "spare" land. However, this potential depends on soil texture and structure, rainfall, and average annual temperatures (Lal, 2004). Nevertheless, Lal (2004) estimated that an increase of one tonne of soil carbon in degraded cropland soils may increase crop yield by 20 to 40 kg per ha for wheat, 10 to 20 kg per ha for maize and 0.5 to one kg per ha for cowpeas. This underlines the complex interdependencies that exist at the nexus of soil carbon, climate change and food/energy production.

Climate change is likely to have a number of effects on nutrient cycling, including notable effects on the global carbon and nitrogen cycles, which have the potential to interact with land degradation processes, through their effects on plant growth via nitrogen availability and $CO_2$ fertilization. Atmospheric concentrations of $CO_2$, methane ($CH_4$) and nitrous oxide ($N_2O$) have all increased since 1750 due to human activity according to IPCC (2013). Twenty-nine per cent of all anthropogenic $CO_2$ emissions have been absorbed by terrestrial ecosystems, primarily forests, leading to increased Net Primary Productivity (see previous section) (IPCC, 2013). Concentrations of $N_2O$ in the atmosphere have been steadily increasing over the last three decades, in addition to elevations in concentrations of other nitrogen compounds (primarily $NO_x$ and $NH_3$). These gases have been implicated in the production of tropospheric ozone, which can impede plant growth (see previous section) (IPCC, 2013). However, where nitrogen is deposited on terrestrial ecosystems, it can increase the productivity of plants, notably forests (IPCC, 2013). Climate warming can increase the rate at which soil organic matter decomposes and the rate of nitrogen mineralization, which can increase nutrient uptake and carbon storage by vegetation, and enhance the productivity of the land for agriculture (IPCC, 2013).

## 5.3 Regulating ecosystem services

Regulating ecosystem services are likely to be exposed and sensitive to a range of potential interactions between climate change and land degradation. Effects of climate change on the regulation of water quality and supplies for agriculture are likely to have a major impact on land degradation processes, leading to land abandonment where it is no longer possible to irrigate crops and water livestock. At the same time, land degradation can contribute towards and exacerbate water quality and supply problems through erosion, which can lead to the sedimentation of dams used for irrigation and the release of nutrients and stored pollutants from historic atmospheric deposition (e.g. heavy metals). Although winter base flow and mean annual stream flow is predicted to increase in most regions under climate change (IPCC, 2013), reduced summer rainfall predicted in some parts of the world may reduce the volume of water in rivers, leading to the concentration of pollutants in stream water to levels that may be toxic for use in agriculture (Confalonieri *et al.*, 2007). The livelihood and wider economic consequences of this for irrigated agricultural systems may be significant, as has been illustrated in the case of the Aral Sea in Central Asia (Box 5.1).

---

### BOX 5.1: IRRIGATION IN CENTRAL ASIA AND THE DRYING OF THE ARAL SEA

Identifying which areas might be suitable for irrigation and the most appropriate scale at which irrigation should take place under future climate conditions is an important challenge to contend with, not least because of the experiences and lessons learned from the past. One example of decision making regarding irrigation which has led to severe ecological damage is found in Central Asia.

The Aral Sea is located between southern Uzbekistan and northern Kazakhstan and used to be the world's fourth largest saline lake. The Aral Sea's water comes from the Amu Darya and Syr Darya Rivers, both of which pass through several countries before reaching the Aral. In the 1960s, the area was governed as part of the USSR, at a time when the Soviet government was aiming to increase self-sufficiency in the production of cotton and rice. To increase productivity, it required the expansion of agriculture into new areas, through the development of irrigation systems. The system of large dams, channels, canals and pipes that ensued, including an 850 mile long central canal, resulted in increased extraction from both rivers in order to irrigate millions of hectares of land. The environmental and social impacts of this water diversion in the decades that followed have been stark. Notable changes include a vast drop in the Aral Sea's water level, receding of the shoreline, increased salinity of the remaining water, and loss of biodiversity within the sea, with fish and other marine life being badly affected. While the irrigated fields initially flourished, the vast areas of monoculture were left vulnerable to pests and diseases, requiring massive applications of pesticides. Estimates suggest that between 1980 and 1992, 72 kg per ha of pesticides were applied in the Aral area, compared with 1.6 kg per ha in the USA. This meant that any runoff that did reach the sea was laden with salts and pesticides from the fields. Drinking water was also contaminated with chemicals, presenting an important hazard to human health. As the water available for irrigation declined, and in combination with changing political and economic governance structures, it left large, dry areas vulnerable to wind erosion. As winds picked up the light, toxic, sandy dust, it was deposited in other parts of the region. The human health impacts resulting from this have been particularly unpleasant. The area sees exceptionally high rates of tuberculosis, anaemia, kidney and liver diseases, respiratory infections, allergies and cancer. Higher risks of infertility, pregnancy complications and miscarriage, as well as brain damage and weakened immune systems of future generations are also present. Overall, an estimated 3.5 million people have been affected by the drying of the Aral Sea, with approximately 60,000 fisherfolk losing their livelihoods (Whish-Wilson, 2002).

The unsustainable extraction of water for irrigation has affected the climate as well. Land next to a large water body is usually warmer in winter and cooler in summer than land further away. As the Aral Sea shrunk, the climate became more extreme. Previous to its drying, the Aral Sea mitigated the cold Siberian winds and reduced the summer heat. Changes to the micro-climate of the region that are linked to the drying of the sea have resulted in drier, shorter summers and longer, colder winters, with an associated reduction in growing seasons and in rangeland productivity. The impacts of this have included the restriction of livelihood options for local people and migration away from affected areas in search of alternative employment opportunities.

Another important regulating ecosystem service is that of pollination. The yields of many of the world's most important crops are dependent upon pollinators, representing approximately 35 per cent of global food production (Klein *et al.*, 2007). The estimated value of crop pollination globally is around

€153 billion annually (Gallai *et al.*, 2009). Given the sensitivity of many insects to small changes in temperature, climate change is likely to interact with a number of other processes to negatively impact upon plant-pollinator interactions (Kjøhl *et al.*, 2011), including interactions with invasive species (Memmott and Waser 2002; Bjerknes *et al.*, 2007), pesticide use (Kearns *et al.*, 1998; Kremen *et al.*, 2002), land-use changes such as habitat fragmentation (Steffan-Dewenter and Tscharntke 1999; Mustajarvi *et al.*, 2001; Aguilar *et al.*, 2006) and agricultural intensification (Tscharntke *et al.* 2005; Ricketts *et al.*, 2008). Where these interactions lead to the extinction of wild pollinator species, this could significantly constrain the production of many crops, compounding the effects of land degradation on livelihoods (Box 5.2).

---

## BOX 5.2: POLLINATOR DECLINE

The current body of research on the impacts of climate change on pollinators is dominated by studies on butterflies, and shows that climate changes have already affected distributions. Studies of British bumblebees also highlight a relationship between climatic niches and species decline, supporting negative projections for future bee species richness in Europe. Climate change is thought to affect pollinators at a range of different scales and in a range of different ways – from altered timings of bee activities, evolutionary changes, species level shifts and local and regional species extinctions to alterations in the composition and functioning of pollinator communities (Potts *et al.*, 2010).

Aside from climate change, the ways in which land is used and managed has an important impact on pollinator populations, and many people advocate the minimal use of farm chemicals (pesticides, herbicides), engagement in organic production methods and the use of integrated pest management strategies. The FAO reports that cardamom farmers in India's Western Ghats are using trees to alter the shade in their fields to ensure suitable conditions for pollinators. They are also planting a range of tree species that bloom sequentially in order to provide on-going pollen and nectar at the times of year when the cardamom is not in bloom. This helps to maintain pollinator presence in the area and reduce the number of bees that leave cardamom plantations in the post-bloom season. Other land management strategies that involve the combination of mixed cropping, kitchen gardens and agroforestry can improve bee habitat, while the maintenance of permanent hedgerows and flower-rich field margins can support a wide range of pollinators. It is also increasingly recognized that farmers often benefit from being close to areas of natural vegetation, as these areas can also help to sustain pollinators, so providing landscape scale connectivity is important.

## 5.4 Cultural ecosystem services

There have been few studies of the likely effects of climate change and land degradation on cultural services. The MA (2005) considered trends in what they called cultural identity, cultural heritage, spiritual services, inspirational services, aesthetic services, recreation and tourism. Under the heading of "cultural identity", the MA (2005) discussed the trend towards sedentarizing nomadic groups around the world.

There have been some attempts to reform land tenure or relocate human populations in response to land degradation and climate change, with consequent effects on cultural heritage, identity and traditions. For example, some authors anticipate mass migration in response to climate change in areas affected by sea level rise, or where, due to the combined effects of climate change and land degradation, it is no longer possible to sustain livelihoods (Bardsley and Hugo, 2010). This may give rise to large populations of environmental refugees, who will typically seek but not always find compatible cultural settings in which they can integrate, leading to a potential loss of cultural identity and heritage (Williams, 2008).

Attempts to reform land tenure in response the effects of land degradation and climate may lead to similar cultural impacts. Following Hardin's (1968) conception of the Tragedy of the Commons, there was a belief that communal systems of livestock management were leading to land degradation, and that nationalization or privatization of rangelands could lead to more sustainable management. However, Hardin was in fact referring to open access regimes, which were *de facto* created when national institutions took over the management of rangelands. In many cases, privatization of rangelands had similar effects, with limited improvements in the sustainability of rangeland management (and in some cases worsened degradation) due to the inability to extend forage range during drought after fencing (e.g. see Perkins, 1996; Mulale *et al.*, 2014 for examples from Botswana). The fencing of communal grazing land also significantly limits the capacity of the system to adapt to droughts, which are predicted to become more frequent and severe in many semi-arid rangelands (Reed *et al.*, 2007; Nori *et al.*, 2008; Stringer *et al.*, 2009a). An alternative, more appropriate solution to Hardin's tragedy in many locations may be to revert to common property regimes. Ostrom (1999) shows how (well-designed and well-implemented) common property regimes are more likely to foster innovative solutions to the challenges of land degradation and climate change, whilst preventing further alienation of commonly held rangeland resources by wealthy individuals (Taylor, 2004).

Spiritual benefits may be derived from particular landscapes and landscape features, such as sacred or holy places (e.g. sacred groves, mountains or waterfalls) or species of plants or animals (e.g. used in ceremonies) (MA, 2005). Spiritual benefits may also be derived from the journeys or pilgrimages to these holy sites, through landscapes that are imbued with meaning by the experiences of those who have passed before (Frey, 1998). In some cases, veneration for particular species or places has afforded protection against over-use or degradation, e.g. the protection of endangered plant species including rare herbs and medicinal plants in sacred groves by Meghalaya tribal communities in northeast India, in an otherwise degraded forest environment (Tiwari et al., 1998). Similarly, taboos have led to the retention of Boscia albitrunca (Shepherd's Tree) in degraded semi-arid savanna in southern Africa (Reed et al., 2007). Boscia albitrunca retains forage all year round, and is a valuable asset during drought, providing opportunities to adapt to climate change (Reed et al., 2014).

Kenter et al. (2014) take this a step further to argue for a link between the aesthetic qualities of particular locations or landscape features and spiritual experience, pointing to evidence that aesthetic and spiritual experiences can be co-emergent. They argue that aesthetic valuing of nature overlaps with spiritual values such as reverence and caring, and stand in opposition to the commodification and degradation of the natural environment. Some of these considerations lay behind the development of the European Landscape Convention, which came into force in 2004. Its goal is to protect and reflect "European identity and diversity, [as] the landscape is our living natural and cultural heritage, be it ordinary or outstanding, urban or rural, on land or in water".

Although some studies have found an aesthetic preference for natural environments in good ecological condition, compared to degraded ecosystems (e.g. Ulrich, 1986), it is important to recognize that many of the aesthetic characteristics consistently appreciated by humans have nothing to do with ecological condition, for example depth of view, openness and the presence of water (Tveit et al., 2006). As a result, local populations may come to appreciate the aesthetic value of exotic invasive species and oppose their removal (Genovesi, 2005). Indeed, Everard et al. (2010) point to the overlooked aesthetic value of dune systems, and Nordstrom and Lotstein (1989) go as far as to argue against the restoration of active dune systems (and subsequent exclusion of people) due to the aesthetic value of unvegetated dunes for local populations who appreciate their "shelter, privacy, definition of territory and pleasing color and texture". Although most of this literature relates to coastal dune systems in developed world contexts, it may be possible to generalize to arid dune systems in developing world environments. However, caution is required, as aesthetic perceptions tend to vary significantly between cultures and over time. For example, Western perceptions of wilderness have changed from attitudes of indifference and hostility until the seventeenth century (which still persist in some cultures) to a more romantic notion of wilderness associated with beauty and freedom (Nash, 2014). Within cultures, perceptions of aesthetic value also vary significantly from person to person, depending on cultural norms, and personal experiences and expectations, for example relating to perceptions of a particular landscape's land ownership, land use and cultural history (Gobster et al., 2007). These aesthetic values may or may not correspond to the ecological status of the land. For example, landscapes with densely vegetated patches of woodland tend to support a wider diversity of wildlife habitats, while evidence from the literature on the aesthetic quality of natural environments has repeatedly established that people tend to prefer more open grassy areas with occasional groupings of trees and shrubs (Parsons, 1995). Similarly, proponents of re-wilding argue that extensively grazed systems are ecologically degraded, while agriculturalists view rewilded landscapes as "abandoned" and degraded in terms of their productive potential (Soliva *et al.*, 2008).

## 5.5 Feedbacks between climate change and land degradation in regions affected by DLDD

There is evidence that land degradation contributes towards climate change, while climate change can exacerbate land degradation (Cowie *et al.*, 2011). However, the numerous feedbacks between climate change and land degradation in regions affected by DLDD, which are likely to mediate the interactions and impacts of these two processes on livelihoods, are poorly understood. Principal among these are feedbacks between:

1.  Climate change, land degradation and carbon sequestration and storage in the soils and vegetation of regions affected by DLDD. Local, regional and global climate patterns are strongly affected by land cover conditions. Moreover, land degradation also affects the organic carbon and nitrogen cycles, altering emissions of organic carbon and nitrogen from soils to the atmosphere and as such, affecting the climate itself. Nevertheless, there are still important knowledge gaps regarding the impact of land degradation and specifically soil erosion on the overall carbon budget, especially regarding possible stabilization mechanisms after erosion and increased inputs of organic carbon to the soil (Boix-Fayos *et al.*, 2014).
2.  Feedbacks between climate change and land degradation-induced changes in vegetation cover and type and local and global climate. Locally, this may lead to changes in regional atmospheric circulation, leading to drier conditions at particular times of year. Globally, changes in the reflectivity (or albedo) of land as a result of these changes in vegetation cover may dampen the effects of global climate change.
3.  There may also be feedbacks between the effects of climate change and land degradation on biodiversity and the provision of a range of ecosystem services, which may further exacerbate land degradation and compromise capacities to adapt to climate change and maintain sustainable livelihoods.

Whether directly via the loss of ecosystem service provision, or indirectly through impacts on the climate system, each of these feedbacks has the potential to reduce livelihood opportunities. It is therefore important to unpack these feedbacks in order to better understand threats and opportunities for livelihoods and develop strategies that can enable livelihood adaptation.

### 5.5.1 *Feedbacks with carbon in vegetation and soil*

Greenhouse gas emissions associated with the loss of soil and vegetation are the key mechanism through which land degradation contributes towards climate change

(MA, 2005). This relationship represents a potential feedback loop, where land degradation leads to climate change, which in turn worsens land degradation. Dryland soils alone, which are particularly vulnerable to degradation, contain more than a quarter of all of the organic carbon stores in the world and nearly all the inorganic carbon (MA, 2005). The loss to the atmosphere of organic carbon held in the drylands could therefore have significant consequences for the global climate system (Kirschbaum, 2006; Heimann and Reichstein, 2008). Indeed, it is already estimated that around 300 million tons of carbon are lost to the atmosphere from drylands as a result of land degradation every year; equivalent to around 4 per cent of total global emissions from all sources combined (MA, 2005; Neely et al., 2009). Conversely, restoration of degraded drylands could sequester 12 to 20 PgC globally over 50 years, if a range of Sustainable Land Management (SLM) practices were successfully applied at a global scale (Lal, 2001, 2004; Suleimenov and Thomas, 2007; Thomas, 2008). Thomas (2008) concluded: "these estimates are admittedly crude but highlight the magnitude of the potential benefits that can be achieved globally through a focus on SLM". In reality, sequestration will be limited by low net primary productivity and the limited capacity of many dryland soils for stabilizing organic matter (Kimetu et al., 2009; Neely et al., 2009). Moreover, feedbacks between climate change, land degradation and losses of carbon from soil and vegetation are highly likely, given the number of mechanisms through which soil carbon may be lost to the atmosphere as a result of both climate change and land degradation. Key mechanisms driving this feedback are:

• enhanced soil respiration under climate warming;
• drought-induced wildfire; and
• land use and management practices leading to a reduction in carbon sequestration and storage by soils and vegetation and/or losses of GHGs from soil and vegetation, including wind and water erosion of soils, over-grazing and over-harvesting (including deforestation).

Climate warming of 2.4°C is predicted to increase soil respiration by around 20 per cent under conditions of sufficient moisture (Norby et al., 2007). For example, climate warming has been blamed for a 2 per cent per annum loss of carbon from soils in England and Wales over the last 50 years (Bellamy et al., 2005). Although respiration losses may be lower in reality (e.g. due to acclimatization of the soil microbial community), climate warming is likely to lead to a progressive loss of soil carbon to the atmosphere, where it can contribute towards future climate change. At the same time, soil carbon plays a critical role in the provision of many ecosystem services linked to plant production, which in turn support rural livelihoods. In drylands particularly, soil carbon has been linked to increased resilience to climate variability and change (Cowie et al., 2011). This is in part because soil carbon enhances, amongst others, infiltration of water and the retention of moisture in soils, improving water availability to plants.

Carbon losses from vegetation due to fire under climate change are likely to be most significant in dryland scrublands and forests, which are currently mainly used for livestock grazing (Howden *et al.*, 1999; MA, 2005). An increased incidence and severity of droughts could increase the likelihood of wildfires (Lindner *et al.*, 2010). The likely impacts of drought on forests has been projected for several regions (e.g. the Amazon and Europe; Cox *et al.*, 2004; Schaphoff *et al.*, 2006; Scholze *et al.*, 2006), showing impacts on forest net ecosystem productivity and wildfire risk, with the potential for a positive feedback to climate change through the release of carbon to the atmosphere and influences on regional climate. In some places a lack of fire can have important negative consequences for ecosystem functioning as the seeds of some plants depend on fire to break their dormancy and allow them to grow. However, there is also evidence that wildfires can lead to long-term land degradation, because the removal of vegetation cover makes soils more susceptible to erosion (Hobbs and Norton, 1996). Where the dominant tree species are already at the margins of their climatic range, there is an increased probability under climate change that wildfire may lead to a shift to a new eco-logical state, possibly dominated by grass or shrubs, which may store less carbon than the previous state (Dougill *et al.*, 1999; Suding *et al.*, 2004). Depending on the extent to which the new state can meet land user objectives, this may or may not constitute land degradation.

Land use and management play a significant role in the release of carbon from soils and vegetation. Grazing-induced land degradation has been estimated to emit as much as 100 million tonnes of $CO_2$ per year globally in drylands alone (FAO/ LEAD, 2006). Net carbon losses occur when carbon sequestration by vegetation and soils decreases and/or removal of carbon from vegetation and soils increases (Cowie *et al.*, 2011). For example, deforestation, heavy grazing and conversion from peren-nial to annual plants in rangelands release carbon stored in above-ground and below-ground biomass to the atmosphere. Regular cultivation of soils can reduce soil carbon unless it is replaced, e.g. via the addition of manures, or the retaining of residues (Cowie *et al.*, 2006). These practices come under the broad umbrella of conservation agriculture (Box 5.3).

Maintaining soil carbon using agronomic, vegetative, structural and management measures can help tackle land degradation and enhance resilience to climatic change in a number of ways. Erosion control and soil fertility/productivity increases can both prevent and in theory reverse land degradation (if land degradation is defined as "a reduction in the resource potential of the land" as it is by UNEP, 1997). By reducing the erodibility of soils, organic matter may offer some protection against an increase in the incidence of highly erodible heavy precipitation events, which are predicted to increase in some locations under climate change. By improving the water holding capacity of the soil, organic matter can help protect against the effects of short-term drought. Water retention may also be enhanced through soil and water conservation or water harvesting techniques. Heavier structures such as trenches, dams, bunds and water-spreading weirs may also improve water infiltration, serve as windbreaks and provide protection against sand encroachment.

---

**BOX 5.3: CONSERVATION AGRICULTURE, SOILS AND RESILIENCE TO CLIMATE CHANGE AND LAND DEGRADATION**

By preserving soil carbon, conservation agriculture also contributes towards food security and sustainable livelihoods by enhancing the fertility of agricultural soils. Conservation agriculture is defined by the FAO (2008: online) as "a concept for resource-saving agricultural crop production that strives to achieve acceptable profits together with high and sustained production levels while concurrently conserving the environment". Examples of conservation agriculture techniques include zero-tillage (where soil-disturbing activities are limited only to those necessary to plant seeds and place nutrients – soil surface and residues are retained intact), conservation tillage (which leaves a minimum of 30% of the soil surface covered with crop residues to increase soil organic matter and reduce soil erosion), strip-till (where only the strip where seeds are to be planted is tilled rather than the whole field) and land sparing (where land is taken out of production for conservation purposes as agricultural yields increase) (Green *et al.*, 2005; Kassam *et al.*, 2009). Within this broad approach, a range of soil and water conservation technologies can be identified. WOCAT (2007) defines these as "agronomic, vegetative, structural and/or management measures that prevent and control land degradation and enhance productivity in the field". These fall into four categories: agronomic (e.g. mulching), vegetative (e.g. contour grass strips), structural (e.g. check dams) or management measures (e.g. resting of land).

---

### 5.5.2 Feedbacks with vegetation cover

Both climate change and land degradation can lead to a reduction in biomass and vegetation cover. Climate-induced changes to vegetation cover are typically in response to reduced water availability due to increasing aridity linked to changing precipitation regimes, combined with increased water use due to increased evapotranspiration under higher temperatures. Soils with low vegetation cover are then more susceptible to erosion, with threshold vegetation cover depending on slope, soil type and erosivity of rainfall (for water erosion) or wind-speed (for wind erosion). Although the $CO_2$ fertilization effect is predicted to increase water-use efficiency of plants, water-use efficiency is compromised in degraded land (Snyman, 1998; Prince *et al.*, 1998; Diouf and Lambin, 2001), thus reducing any compensating effect in these areas. Although there is little certainty over likely changes in average wind-speeds under climate change (IPCC, 2007) an increase in the incidence and severity of extreme weather events is predicted, including a highly likely increase in the frequency, intensity and/or amount of heavy precipitation (especially at mid-latitudes), a highly likely increase in the frequency and/or duration of extremely hot weather (or "heat waves") and likely increases in intensity and/or duration of drought (IPCC, 2013).

Human-induced changes to vegetation cover are typically due to over-exploitation of resources, linked, for example, to over-grazing or over-cultivation (sometimes resulting from reduced fallow periods). Together, increases in extreme weather events and human-induced land degradation processes are likely to lead to a loss of vegetation cover across many parts of the world. Thomas (2008) suggests that this combination of effects might be felt particularly acutely where annual rainfall lies between 500–750 mm, because extreme weather events are likely in these regions, and vegetation cover is unlikely to be sufficient to protect soils from the effects of these events. Soil loss may lead to further land degradation, where this leads to a loss of ecosystem resilience, increasing the likelihood that the system will be unable to retain its critical processes and functions after the extreme event (Thomas, 2008).

There are two types of feedback linked to this. First, these changes may alter regional climates due to changes in dust fluxes (Hardy, 2003; Prospero and Lamb, 2003; Lioubimtseva and Adams, 2004) and high-pressure circulation anomalies that can result in drier conditions at particular times of year (McGuffie *et al.*, 1995; Sud *et al.*, 1996; Archer and Tadross, 2009). The second type of feedback relates to increase in the reflectivity or albedo of the land surface when vegetation cover is removed, leading to possible effects on surface evapotranspiration and heat and moisture fluxes. This in turn may affect local, regional and possibly global, atmospheric circulation, leading to a negative feedback to the climate system (Thomas, 2008). However, the opposite is also possible. Increases in vegetation cover can help to restore ecosystem functions and services in degraded areas, with positive implications for both climate change and progress towards LDN (Box 5.4).

---

## BOX 5.4: TREES FOR LAND DEGRADATION NEUTRALITY AND CLIMATE RESILIENCE

Whether via planting or natural regeneration, trees have the potential to restore ecosystem functions and services to degraded land and increase resilience to climate change in a number of ways. Most obviously, forests that have been cleared (e.g. for agriculture) or degraded (e.g. via selective logging) can regain functions and services through reforestation. However, trees are used to tackle land degradation in a number of other contexts. For example, land reclamation via afforestation may be the only viable option for restoring some levels of biodiversity and ecosystem services in former coal or bauxite mining operations, where soil removal or toxic substrata limit the ability of native vegetation to grow (Prach *et al.*, 2007). In areas with degraded soils, rehabilitation through planting of carefully selected native trees can improve soil fertility and restore productive agricultural use (Chazdon *et al.*, 2008). For example, in the Shinyanga region of Tanzania, large areas of dense acacia and miombo woodland had been cleared by 1985, creating an unproductive grassland that was of limited benefit to pastoralists. Over 18 years, the HASHI program helped local people from 833 villages to restore 350,000 ha of woodland through traditional pastoralist practices (Monela *et al.*, 2004). Reforestation has the capacity to improve the production of food and fibre from degraded land, whilst in some cases also benefiting biodiversity. Of course, where tree cover becomes permanent, the sequestration and subsequent storage of carbon in tree biomass and soil organic matter also has the potential to mitigate climate change. Where this is part of a strategy to protect and enhance water catchments, such schemes can also help to increase the adaptive capacity of communities to extreme weather events, for example moderating peak flows during heavy rainfall events and reducing the likelihood of flash floods.

Agroforestry practices also have significant potential to both tackle land degradation and facilitate climate change mitigation and adaptation. As such, this is one of a number of techniques that may be considered under the broad heading of "climate smart agriculture". IPCC (2000) identified agroforestry as the land use strategy with most mitigation potential, simply due to the vast area of agricultural land over which it could potentially be applied. However, a number of important barriers prevent the adoption of agroforestry on these sorts of scales (Reed, 2007). Agroforestry nevertheless still has significant potential to tackle land degradation and climate change, particularly in smallholder rain-fed agriculture systems where there are few other available options.

Agroforestry can increase resilience to climate change by enhancing the diversity of agricultural products from a farm enterprise (where multi-purpose trees are planted). More diversified farming systems are less likely to suffer from climate-related shocks and more likely to maintain the ability of farmers to adapt to changing conditions (Verchot *et al.*, 2007). Techniques, such as intercropping with leguminous woody species, can help tackle land degradation by allowing access to nutrients deeper in the soil profile, providing resilience to the system during drought. Trees also have the capacity to increase soil porosity and reduce runoff. Combined with increased soil cover, this can lead to increased water infiltration and retention in the soil profile, which can reduce moisture stress during low rainfall years. The same combination of factors also reduces soil erosion. Tree-based systems have the potential to increase the production of crops and livestock under climate change. For example, trees can shelter crops from the effects of wind erosion and storm damage, and shade livestock during heat waves. Depending on the species planted, trees can provide an important buffer to drought through the provision of an all-year round fodder. Thus, integrating trees with agricultural production systems may buffer against a range of risks associated with both land degradation and climate change.

### 5.5.3 *Feedbacks with biodiversity*

Climate change and land degradation have similar effects on biodiversity, leading overall to the simplification of ecosystems and an increased abundance of generalist species at the expense of specialists (Clavel *et al.*, 2010). They can also lead to the loss of genetic diversity. Indeed, recent literature includes loss of biodiversity as a form of land degradation (Haines-Young and Potschin, 2010). Climate change will have a number of indirect impacts on biodiversity, for example, shifts in the timing and success of reproduction (Forchhammer *et al.*, 1998; Crick and Sparks, 1999; Winkler *et al.*, 2002), changes in the availability and suitability of habitats and niches (Visser and Both, 2005), changes in the way species use habitats, e.g. nest and shelter site selection (Telemeco *et al.*, 2009) and changes in survival rates (Chamaille-Jammes *et al.*, 2006). Climate change will directly affect the distribution of species, as the location of climatic zones to which they are adapted generally move polewards and towards higher altitudes (Meynecke, 2004; Penman *et al.*, 2010). This shift in climatic zones will also mean there is likely to be a declining number of habitats and specialist niches, leading to the replacement of specialist species with generalists (MA, 2005). The $CO_2$ fertilization effect is likely to lead to changes in plant species composition, for example favouring nitrogen-fixing and C4 species (MA, 2005; Thomas, 2008). In areas made more susceptible to fire due to climate change and land degradation, species composition is likely to shift towards pyrophytic species, capable of withstanding fire, leading to an overall reduction in biodiversity (Neilson *et al.*, 1998). The mix of species likely to gain a competitive advantage will therefore depend on the assemblage of species exposed to climate change in any given location, so it is difficult to predict how these changes in diversity might affect the provision of ecosystem services, and hence impact upon livelihoods. Canziani *et al.* (1998) suggested dryland species might be less sensitive to climate change because they are already well adapted to climate extremes (see Box 5.5). However, it is clear that this poorly understood feedback is likely to present significant challenges to those whose livelihoods depend on biodiversity (MA, 2005).

It is further important to remember that not all biodiversity is above ground, and that the soil harbours a rich and varied selection of micro-organisms. In drylands, these micro-organisms can often take the form of biological soil crusts (BSC) that cover the top few centimeters of soils. These crusts are composed of complex communities of algae, bacteria, microfungi, lichens and bryophytes. They stabilize the soil and reduce erosion, play a preparatory role in facilitating the colonization of higher plants, and regulate gas fluxes of nitrogen and carbon between the Earth and the atmosphere. Despite these important roles in underpinning the delivery of a range of regulatory and supporting ecosystem services, BSCs and soil microbial communities more generally are rarely given any consideration in conservation planning and assessment. Recent research advances linked to rDNA analysis have hugely improved the speed of analysis such that it has provided new insights into the phylogenetic diversity of soil microbial communities, revealing microbial

---

## BOX 5.5: ESCAPERS, EVADERS, RESISTERS AND ENDURERS

Dryland plant and animal species often exhibit important traits that allow them to survive in water-limited parts of the world. Some plants and animals escape into seeds, eggs or the larval stage until there is sufficient moisture. Other species evade. For example, some plants have very deep or very wide-spread roots enabling them to tap into groundwater that other plants are unable to reach, while reptiles evade extreme heat by hiding themselves underneath the ground surface. A further group of species resist drought through their water storage mechanisms. For instance, camels are well adapted to minimize water losses, while cacti are able to store water in their roots and trunks. A fourth group can endure extreme conditions such as drought by going dormant. Indeed, many architects are learning from these specialized characteristics in order to design buildings that utilize similar adaptive mechanisms, e.g. enabling moisture collection.

Modified from SBSTTA (1999)

---

species that we would not otherwise know about through the use of traditional culturing techniques.

## 5.6 Synthesis

This chapter has considered the likely exposure and sensitivity of supporting, regulating and cultural ecosystem services to climate change and land degradation processes in regions affected by DLDD. Effects on supporting services may include decreased rates of soil formation (e.g. via increased soil organic matter decomposition rates), and increased rates of soil erosion (e.g. via increasingly erosive rainfall). These changes may then interact with changes in temperature, solar radiation and atmospheric $CO_2$ concentrations to affect plant biomass production and reduce vegetation cover, which could in turn further expose soils to erosion. Examples of regulating services likely to be exposed and sensitive to the combined effects of climate change and land degradation include pollination services (e.g. via effects of climate change on pollinator species populations and changes in land use and management in response to land degradation), and changes in temperature effects on the regulation of water quality and supply for agriculture (e.g. via changes in precipitation and erosion processes). Examples of vulnerable cultural ecosystem services include effects on cultural identity and heritage arising from attempts to reform land tenure or relocate human populations in response to land degradation and climate change. Spiritual benefits linked to particular landscapes or landscape features may be threatened by land degradation and climate change, for example, if these processes compromise their aesthetic quality. However, the aesthetic value placed on landscapes is not necessarily linked to their ecological status.

The chapter concludes by considering likely feedbacks between climate change and land degradation in regions affected by DLDD, which are likely to have implications across a wide range of ecosystem services. There is evidence that land degradation contributes towards climate change, while climate change can exacerbate land degradation, but the numerous feedbacks between climate change and land degradation are poorly understood. We identify three feedbacks that may be of particular concern. First, feedbacks may occur between climate change, land degradation and carbon sequestration and storage in the soils and vegetation of regions affected by DLDD. Second, there may be feedbacks between climate change and land degradation-induced changes in vegetation cover and type and local and global climate. Locally, this may lead to changes in regional atmospheric circulation, leading to drier conditions at particular times of year. Globally, alterations in the land's albedo as a result of these changes in vegetation cover may dampen the effects of global climate change. And finally, there may also be feedbacks between the effects of climate change and land degradation on biodiversity and the provision of a range of ecosystem services, which may further exacerbate land degradation and compromise capacities to adapt to climate change and maintain sustainable livelihoods.

This and the previous chapter have outlined the likely exposure and sensitivity of different systems and ecosystem services to the combined effects of climate change and land degradation. Moving to the final elements of the conceptual framework outlined in Chapter 3, any assessment of vulnerability to climate change and land degradation must consider the adaptive capacity of the ecosystems and human populations under consideration. This is the topic of the next chapter.

# References

Aguilar, R., Ashworth, L., Galetto, L., Aizen, M.A. 2006. Plant reproductive susceptibility to habitat fragmentation: review and synthesis through a meta-analysis. *Ecological Letters* 9: 968–980.

Archer, E.R., Tadross, M.A. 2009. Climate change and desertification in South Africa – science and response. *African Journal of Range & Forage Science* 26: 127–131.

Bardsley, D.K., Hugo, G.J. 2010. Migration and climate change: examining thresholds of change to guide effective adaptation decision-making. *Population and Environment* 32 (2–3): 238–262.

Bellamy, P.H., Loveland, P.J., Bradley, R.I., Lark, R.M., Kirk, G.J.D. 2005. Carbon losses from all soils across England and Wales 1978–2003. *Nature* 437: 245–248.

Bernoux M., Chevallier T. 2014. Carbon in dryland soils. Multiple essential functions. *Les dossiers thématiques du CSFD*. N°10. June 2014. CSFD/Agropolis International: Montpellier, France, 40 pp.

Bjerknes, A.L., Totland, O., Hegland, S.J., Nielsen, A. 2007. Do alien plant invasions really affect pollination success in native plant species? *Biological Conservation* 138: 1–12.

Boix-Fayos, C., Nadeu, E., Quiñonero, J.M., Martínez-Mena, M., Almagro, M., de Vente, J. 2014. Coupling sediment flow-paths with organic carbon dynamics across a Mediterranean catchment. *Hydrology and Earth System Sciences* 11: 5007–5036.

Bond-Lamberty, B., Wang, C.K., Gower, S.T. 2004. A global relationship between the heterotrophic and autotrophic components of soil respiration? *Global Change Biology* 10: 1756–1766.

Bond-Lamberty, B., Thomson, A. 2010. Temperature-associated increases in the global soil respiration record. *Nature* 464: 579–582.

Canziani, O.F., Díaz, S., Calvo, E., Campos, M., Carcavallo, R., *et al.* 1998. Latin America. In: *The Regional Impacts of Climate Change: An assessment of vulnerability. Special Report of IPCC Working Group II*, Watson, R.T., Zinyowera, M.C., Moss, R.H. (eds). Intergovernmental Panel on Climate Change. Cambridge University Press: Cambridge, United Kingdom and New York, NY, USA, pp. 187–230.

Cerling T.E. 1984. The stable isotopic composition of modern soil carbonate and its relationship to climate. *Earth and Planetary Science Letters* 71: 229–240.

Chamaille-Jammes, S., Massot, M., Aragon, P., Clobert, J. 2006. Global warming and positive fitness response in mountain populations of common lizards Lacerta vivipara. *Global Change Biology* 12: 392–402.

Chazdon, R.L. 2008. Beyond deforestation: restoring forests and ecosystem services on degraded lands. *Science* 320 (5882): 1458–1460.

Clavel, J., Julliard, R., Devictor, V. 2010. Worldwide decline of specialist species: toward a global functional homogenization? *Frontiers in Ecology and the Environment* 9: 222–228.

Confalonieri, U., Menne, B., Akhtar, R., Ebi, K.L., Hauengue, M., Kovats, R.S., Revich, B., Woodward, A. 2007. Human health. In: *Climate Change 2007: Impacts, adaptation and vulnerability. Contribution of Working Group II to the Fourth Assessment Report of the Intergovernmental Panel on Climate Change*. Cambridge University Press: Cambridge, UK, pp. 391–431.

Council of Europe. 2004. European Landscape Convention. Available online at: http://www.coe.int/en/web/landscape

Cowie, A.L., Smith, P., Johnson, D. 2006. Does soil carbon loss in biomass production systems negate the greenhouse benefits of bioenergy? *Mitigation and Adaptation Strategies for Global Change* 11: 979–1002.

Cowie, A.L., Penman, T.D., Gorissen, L., Winslow, M.D., Lehmann, J., Tyrrell, T.D., Twomlow, S., Wilkes, A., Lal, R., Jones, J.W., Paulsch, A., Kellner, K., Akhtar-Schuster, M. 2011. Towards sustainable land management in the drylands: scientific connections in monitoring and assessing dryland degradation, climate change and biodiversity. *Land Degradation & Development* 22: 248–260.

Cox, P.M., Betts, R.A., Collins, M., Harris, P.P., Huntingford, C., Jones, C.D. 2004. Amazonian forest dieback under climate-carbon cycle projections for the 21st century. *Theoretical Applications of Climatology* 78: 137–156.

Crick, H.Q.P., Sparks, T.H. 1999. Climate change related to egg-laying trends. *Nature* 399: 423–424.

Diouf, A., Lambin, E.F. 2001. Monitoring land-cover changes in semi-arid regions: remote sensing data and field observations in the Ferlo, Senegal. *Journal of Arid Environments* 48: 129–148.

Dougill, A.J., Thomas, D.S.G., Heathwaite, A.L. 1999. Environmental change in the Kalahari: integrated land degradation studies for non-equilibrium dryland environments. *Annals of the Association of American Geographers* 89: 420–442.

Dougill, A.J., Thomas, A.D. 2002. Nebkha dunes in the Molopo Basin, South Africa and Botswana: formation controls and their validity as indicators of soil degradation. *Journal of Arid Environments* 50: 413–428.

Everard, M., Jones, L., Watts, B. 2010. Have we neglected the societal importance of sand dunes? An ecosystem services perspective. *Aquatic Conservation: Marine and Freshwater Ecosystems* 20(4): 476–487.

FAO/LEAD. 2006. *Livestock's Long Shadow. Environmental Issues and Options.* FAO: Rome.

FAO. 2008. What is conservation agriculture? In: Conservation agriculture website of FAO, www.fao.org/ag/ca/1a.html.

Fischlin, A., Midgley, G.F., Price, J.T., Leemans, R., Gopal, B., Turley, C., Rounsevell, M.D.A., Dube, O.P., Tarazona, J., Velichko, A.A. 2007. Ecosystems, their properties, goods, and services. In: *Climate Change 2007: Impacts, adaptation and vulnerability. Contribution of Working Group II to the Fourth Assessment Report of the Intergovernmental Panel on Climate Change,* Parry, M.L., Canziani, O.G., Palutikof, J.P., van der Linden, P.J., Hanson, C.E. (eds). Cambridge University Press: Cambridge, pp. 211–272.

Forchhammer, M.C., Post, E., Stenseth, N.C. 1998. Breeding phenology and climate change. *Nature* 391: 29–30.

Frey, N.L. 1998. *Pilgrim Stories: On and off the road to Santiago.* University of California Press: Berkeley.

Gallai, N., Salles, J.M., Settele, J., Vaissiere, B.E. 2009. Economic valuation of the vulnerability of world agriculture confronted with pollinator decline. *Ecological Economics* 68: 810–821.

Genovesi, P. 2005. Eradications of invasive alien species in Europe: a review. In: *Issues in Bioinvasion Science. Springer Netherlands,* pp. 127–133.

Gobster, P.H., Nassauer, J.I., Daniel, T.C., Fry, G. 2007. The shared landscape: what does aesthetics have to do with ecology? *Landscape ecology* 22(7): 959–972.

Green, R.E., Cornell, S.J., Scharlemann, J.P., Balmford, A. 2005. Farming and the fate of wild nature. *Science* 307(5709): 550–555.

Haines-Young, R., Potschin, M. 2010. The links between biodiversity, ecosystem services and human well-being. *Ecosystem Ecology: A New Synthesis,* pp. 110–139.

Hamdi, S., Chevallier, T., Ben Aïssa N., Ben Hammouda M., Gallali T., Chotte J.L., Bernoux M. 2011. Short-term temperature dependence of heterotrophic soil respiration after one-month of pre-incubation at different temperatures. *Soil Biology and Biochemistry* 43: 1752–1758.

Hamdi, S., Moyano, F., Sall, S., Bernoux, M., Chevallier, T. 2013. Synthesis analysis of the temperature sensitivity of soil respiration from laboratory studies in relation to incubation methods and soil conditions. *Soil Biology and Biochemistry* 58: 115–126.

Hardin, G. 1968. The tragedy of the commons. *Science* 162, 1243–1248.

Hardy, J.T. 2003. *Climate Change Causes, Effects and Solutions.* Wiley: Chichester.

Heimann, M., Reichstein, M. 2008. Terrestrial ecosystem carbon dynamics and climate feedbacks. *Nature* 451: 289–292.

Hobbs, R. J., Norton, D.A. 1996. Towards a conceptual framework for restoration ecology. *Restoration Ecology* 4: 93–110.

Howden, S.M., Reyenga, P.J., Meinke, H., McKeon, G.M. 1999. Integrated global change impact assessment on Australian terrestrial ecosystems: overview report. *Working Paper Series 99/14, CSIRO Wildlife and Ecology.* Canberra, Australia, 51 pp.

IPCC. 2000. *Land Use, Land-use Change and Forestry. A Special Report of the IPCC.* Cambridge University Press: Cambridge.

IPCC. 2007. Climate Change 2007: The Physical Science Basis. In: *Contribution of Working Group I to the fourth assessment report of the intergovernmental panel on climate change,* Solomon, S., Qin, H.D., Manning, M., Chen, Z., Marquis, M., Averyt, K.B., Tignor, M., Miller, H.L. (eds). Cambridge University Press: Cambridge, United Kingdom and New York, NY, USA, 996 pp.

IPCC 2013. Climate Change 2013. The Physical Science Basis. In: *Contribution of Working Group I to the fifth assessment report of the intergovernmental panel on climate change*, Stocker, T.F., Qin, D., Plattner, G.K., Tignor, M., Allen, S.K., Boschung, J., Nauels, A., Xia, Y., Bex, V., Midgley, P.M. (eds). Cambridge University Press: Cambridge, United Kingdom and New York, NY, USA, 1535 pp.

Kassam, A., Friedrich, T., Shaxson, F., Pretty, J. 2009. The spread of conservation agriculture: justification, sustainability and uptake. *International Journal of Agricultural Sustainability* 7(4): 292–320.

Kearns, C.A., Inouye, D.W., Waser, N.M. 1998. Endangered mutualisms: the conservation of plant-pollinator interactions. *Annual Review of Ecology, Evolution and Systematics* 29: 83–112.

Kenter, J.O., Reed, M.S., Irvine, K.N., O'Brien, E., Brady, E., Bryce, R., Christie, M., Church, A., Cooper, N., Davies, A., Hockley, N., Fazey, I., Jobstvogt, N., Molloy, C., Orchard-Webb, J., Ravenscroft, N., Ryan, M., Watson, V. 2014. *UK National Ecosystem Assessment Follow-on Phase, Technical Report: Shared, plural and cultural values of ecosystems.* UNEP-WCMC, Cambridge.

Kimetu, J.M., Lehmann, J., Kinyangi, J.M., Cheng, C.H., Thies, J., Mugendi, D.N., Pell, A. 2009. Soil organic C stabilization and thresholds in C saturation. *Soil Biology and Biochemistry* 41: 2100–2104.

Kirschbaum, M.U.F. 2006. The temperature dependence of organic-matter decomposition – still a topic of debate. *Soil Biology and Biochemistry* 38: 2510–2518.

Kjøhl, M., Nielsen, A., Stenseth, N.C. 2011. Potential effects of climate change on pollination. FAO: Rome. Available online at: www.fao.org/docrep/014/i2242e/i2242e.pdf.

Klein, A.M., Vaissiere, B.E., Cane, J.H., Steffan-Dewenter, I., Cunningham, S.A., Kremen, C., Tscharntke, T. 2007. Importance of pollinators in changing landscapes for world crops. *Proceedings of the Royal Society of London B* 274: 303–313.

Kremen, C., Williams, N.M., Thorp, R.W. 2002. Crop pollination from native bees at risk from agricultural intensification. *Proceedings of the National Academy of Sciences USA* 99: 16812–16816.

Lal, R., Bruce J.P. 1999. The potential of world cropland soils to sequester C and mitigate the greenhouse effect. *Environmental Science & Policy* 2(2): 177–185.

Lal, R. 2001. The potential of soils of the tropics to sequester carbon and mitigate the greenhouse effect. *Advances in Agronomy* 76: 1–30.

Lal, R. 2004. Carbon sequestration in dryland ecosystems. *Environmental Management* 33: 528–544.

Lindner, M., Maroschek, M., Netherer, S., Kremer, A., Barbati, A., Garcia-Gonzalo, J., Seidl, R., Delzon, S., Corona, P., Kolstrom, M., Lexer, M.J., Marchetti, M. 2010. Climate change impacts, adaptive capacity and vulnerability of European forest ecosystems. *Forest Ecology and Management* 259: 698–709.

Lioubimtseva, E., Adams, J.M. 2004. Possible implications of increased carbon dioxide levels and climate change for desert ecosystems. *Environment Management* 33: S388–S404.

Luo, Y.W.S., Hui, D., Wallace, L.L. 2001. Acclimatization of soil respiration to warming in a tall grass prairie. *Nature* 413: 622–625.

MA (Millennium Ecosystem Assessment). 2005. *Ecosystems and Human Well-being: Current state and trends assessment.* Island Press: Washington, D.C, USA.

McGuffie, K.A., Henderson-Sellers, A., Zhang, H., Durbidge, T.B., Pitman, A.J. 1995. Global climate sensitivity to tropical deforestation. *Global Planetary Change* 10: 97–128.

Memmott, J., Waser, N.M. 2002. Integration of alien plants into a native flower-pollinator visitation web. *Proceedings of the Royal Society of London B* 269: 2395–2399.

Meynecke, J.-O. 2004. Effects of global climate change on geographic distributions of vertebrates in North Queensland. *Ecological Modelling* 174: 347–357.

Monela, G., Chamshama, S., Mwaipopo, R., Gamassa, D. A. 2004. *Study on the Social, Economic and Environmental Impacts of Forest Landscape Restoration in Shinyanga Region, Tanzania.* The United Republic of Tanzania Ministry of Natural Resources and Tourism, Forestry, and Beekeeping Division: Dar-es-Salaam, Tanzania, and IUCN, The World Conservation Union, Eastern Africa Regional Office: Nairobi, Kenya.

Mulale, K., Chanda, R., Perkins, J.S., Magole, L., Sebego, R.J., Atlhopheng, J.R., Mphinyane, W., Reed, M.S. 2014. Formal institutions and their role in sustainable land management in Boteti, Botswana. *Land Degradation & Development* 25: 80–91.

Mustajarvi, K., Siikamaki, P., Rytkonen, S., Lammi, A. 2001. Consequences of plant population size and density for plant-pollinator interactions and plant performance. *Journal of Ecology* 89: 80–87.

Nash, R.F. 2014. *Wilderness and the American Mind.* Yale University Press: New Haven, CT.

Nearing, M.A. 2001. Potential changes in rainfall erosivity in the US with climate change during the 21st century. *Journal of Soil and Water Conservation* 56: 229–232.

Nearing, M.A., Pruski, F.F., O'Neal, M.R. 2004. Expected climate change impacts on soil erosion rates: a review. *Journal of Soil and Water Conservation* 59: 43–50.

Neely, C., Bunning, S., Wilkes, A. 2009. Review of evidence on dryland pastoral systems and climate change: implications and opportunities for mitigation and adaptation. *Land and Water Discussion Paper 8.* FAO: Rome.

Neilson, R.P., Prentice, I.C., Smith, B., Kittel, T., Viner, D. 1998. Simulated changes in vegetation distribution under global warming. In: *Regional Impacts of Climatic Change: An assessment of vulnerability. A Special Report of the Intergovernmental Panel on Climate Change (IPCC) Working Group II,* Watson, R.T., Zinyowera, M.C., Moss, R.H., Dokken, D.J., (eds). Cambridge University Press: Annex C, 439–456.

Norby, R.J., Rustad, L.E., Dukes, J.S., Ojima, D.S., Parton, W.J., Del Grosso, S.J., McMurtrie, R.E., Pepper, D.P. 2007. Ecosystem responses to warming and interacting global change factors. In: *Terrestrial Ecosystems in a Changing World,* Canadell, J.G., Pataki. D.E., Pitelka, L.F. (eds). Springer-Verlag: Berlin, 45–58.

Nordstrom, K.F., Lotstein, E.L. 1989. Perspectives on resource use of dynamic coastal dunes. *Geographical Review* 1–12.

Nori, M., Taylor, M., Sensi, A. 2008. Browsing on fences: pastoral land rights, livelihoods and adaptation to climate change. No. 148. International Institute for Environment and Development, London, UK.

Ostrom, E.M. 1999. Institutional rational choice: an assessment of the Institutional Analysis and Development Framework. In: *Theories of the Policy Process:* 35–71; Sabatier, P.A. (ed.). Westview Press: Boulder, CO, USA.

Parsons, R. 1995. Conflict between ecological sustainability and environmental aesthetics: conundrum, canärd or curiosity. *Landscape and Urban Planning* 32(3): 227–244.

Penman, T.D., Pike, D.A., Webb, J.K., Shine, R. 2010. Predicting the impact of climate change on Australia's most endangered snake, *Hoplocephalus bungaroides. Diversity and Distributions* 16: 109–118.

Perkins, J.S. 1996. Botswana: fencing out the equity issue. Cattleposts and cattle ranching in the Kalahari Desert. *Journal of Arid Environments* 33: 503–517.

Potts, S.G., Biesmeijer, J.C., Kremen, C., Neumann, P., Schweiger, O. and Kunin, W.E. 2010. Global pollinator declines: trends, impacts and drivers. *Trends in Ecology & Evolution* 25(6): 345–353

Prach, K., Marrs, R., Pysek, P., van Diggelen, R. 2007. Manipulation of Succession. In: *Linking Restoration and Ecological Succession.* Walker, L.R., Walker, J., Hobbs, R.J. (eds). Springer: New York, pp. 121–149.

Prince, S.D., De Colstoun, E.B., Kravitz, L.L. 1998. Evidence from rain-use efficiencies does not indicate extensive Sahelian desertification. *Global Change Biology* 4: 359–374.

Prospero, J.M., Lamb, P.J. 2003. African droughts and dust transport to the Caribbean: climate change implications. *Science* 302: 1024–1027.

Pruski, F.F., Nearing, M.A. 2002. Climate-induced changes in erosion during the 21st century for eight US locations. *Water Resources Research* 38: 34-134-11

Reed, M.S. 2007. Participatory technology development for agroforestry extension: an innovation-decision approach. *African Journal of Agricultural Research* 2: 334–341.

Reed, M.S., Dougill A.J., Taylor M.J. 2007. Integrating local and scientific knowledge for adaptation to land degradation: Kalahari rangeland management options. *Land Degradation & Development* 18: 249–268.

Reed, M.S., Stringer, L.C., Fazey, I., Evely, A.C., Kruijsen, J. 2014. Five principles for the practice of knowledge exchange in environmental management. *Journal of Environmental Management* 146: 337–345.

Reed, M.S., Stringer, L.C., Dougill, A.J., Perkins, J.S., Atlhopheng, J.R., Mulale, K., Favretto, N. 2015. Reorienting land degradation towards sustainable land management: linking sustainable livelihoods with ecosystem services in rangeland systems. *Journal of Environmental Management* 151: 472–485.

Ricketts, T.H., Regetz, J., Steffan-Dewenter, I., Cunningham, S.A., Kremen, C., Bogdanski, A., Gemmill-Herren, B., Greenleaf, S.S., Klein, A.M., Mayfield, M.M., Morandin, L.A., Ochieng, A., Viana, B.F. 2008. Landscape effects on crop pollination services: are there general patterns? *Ecological Letters* 11: 1121–1121.

Rillig, M.C., Wright, S.F., Shaw, M.R., Field, C.B. 2002. Artificial climate warming positively affects arbuscular mycorrhizae but decreases soil aggregate water stability in an annual grassland. *Oikos* 97: 52–58.

Schlesinger W.H. 1982. Carbon storage in the caliche of arid soils: a case study from Arizona. *Soil Science* 133: 247–255.

Schlesinger, W.H., Reynolds, J.F., Cunningham, G.L., Huenneke, L.F., Wesely, M.J., Virginia, R.A., Whitford, W.G. 1990. Biological feedbacks in global desertification. *Science* 24: 1043–1048.

SBSTTA, 1999. Biological diversity of drylands, arid, semi-arid, savannah, grassland and Mediterranean ecosystems. Draft recommendations to COP5. Montreal, Canada, 20 pp.

Schaphoff, S., Lucht, W., Gerten, D., Sitch, S., Cramer, W., Prentice, I.C. 2006. Terrestrial biosphere carbon storage under alternative climate projections. *Climatic Change* 73: 97–122.

Scholze, M., Knorr, W., Arnell, N.W., Prentice, I.C. 2006. A climate change risk analysis for world ecosystems. *Proceedings of the National Academy of Science USA* 103: 13116–13120.

Snyman, H.A. 1998. Dynamics and sustainable utilization of rangeland ecosystems in arid and semi-arid climates of southern Africa. *Journal of Arid Environments* 39: 645–666.

Soliva, R., Rønningen, K., Bella, I., Bezak, P., Cooper, T., Flø, B.E., Potter, C. 2008. Envisioning upland futures: stakeholder responses to scenarios for Europe's mountain landscapes. *Journal of Rural Studies* 24(1): 56–71.

Steffan-Dewenter I., Tscharntke, T. 1999. Effects of habitat isolation on pollinator communities and seed set. *Oecologia* 121: 432–440.

Stringer L.C., Dyer, J., Reed, M., Dougill, A., Twyman, C., Mkwambisi, D. 2009. Adaptations to climate change, drought and desertification: insights to enhance policy in southern Africa. *Environmental Science and Policy* 12: 748–765.

Stringer, L.C., Scrieciu, S.S., Reed, M.S. 2009. Biodiversity, land degradation, and climate change: Participatory planning in Romania. *Applied Geography* 29: 77–90.

Sud, Y.C., Walker, G.K., Kim, J.-H., Liston, G.E., Sellers, P.J., Lau, K.-M. 1996. Biogeophysical effects of a tropical deforestation scenario: a GCM simulation study. *Journal of Climate* 16: 135–178.

Suding, K.N., Gross, K.L., Houseman, G.R. 2004. Alternative states and positive feedbacks in restoration ecology. *Trends in Ecology & Evolution* 19: 46–53.

Suleimenov, M., Thomas, R.J. 2007. Central Asia: ecosystems and carbon sequestration challenges. In: *Climate Change and Terrestrial Carbon Sequestration in Central Asia*, Lal, R., Suleimenov, M., Stewart, B.A., Hansen, D.O., Doraiswamy, P. (eds). Taylor & Francis: London, 165–176.

Taylor, M. 2004. The past and future of San land rights in Botswana. In: *Indigenous Peoples' Rights in Southern Africa*, Hitchcock, R., Vinding, D. (eds). International Working Group for Indigenous Affairs (IWGIA): Copenhagen.

Telemeco, R.S., Elphick, M.J., Shinem R. 2009. Nesting lizards (*Bassiana duperreyi*) compensate partly, but not completely, for climate change. *Ecology* 90: 17–22.

Thomas, R.J. 2008. 10th Anniversary review: Addressing land degradation and climate change in dryland agroecosystems through sustainable land management. *Journal of Environmental Monitoring* 10: 595–603.

Tiwari, B.K., Barik, S.K., Tripathi, R.S. 1998. Biodiversity value, status, and strategies for conservation of sacred groves of Meghalaya, India. *Ecosystem Health* 4: 20–32.

Tscharntke, T, Klein, A.M., Kruess, A, Steffan-Dewenter, I., Thies, C. 2005. Landscape perspectives on agricultural intensification and biodiversity-ecosystem service management. *Ecological Letters* 8: 857–874.

Tveit, M., Ode, Å., Fry, G. 2006. Key concepts in a framework for analysing visual landscape character. *Landscape Research* 31: 229–255.

Ulrich, R. 1986. Human response to vegetation and landscapes. *Landscape and Urban Planning* 13: 29–44.

UNEP. 1997. *World Atlas of Desertification*, 2nd edition. Middleton, N., Thomas, D. (eds). London: Arnold.

Verchot, L.V., Van Noordwijk, M, Kandji, S., Tomich, T.P., Ong, C.K., Albrecht, A., Mackensen, J., Bantilan, C., Anupama, K.V., Palm, C.A. 2007. Climate change: linking adaptation and mitigation through agroforestry. *Mitigation and Adaptation Strategies for Global Change* 12: 901–918.

Visser, M.E., Both, C. 2005. Shifts in phenology due to global climate change: the need for a yardstick. *Proceedings of the Royal Society of London Series B-Biological Sciences* 272: 2561–2569.

Warren, A.S. 2002. Land degradation is contextual. *Land Degradation and Development* 13: 449–459.

Whish-Wilson P. 2002. The Aral Sea environmental health crisis. *Journal of Rural and Remote Environmental Health* 1(2): 29–34

Williams, A. 2008. Turning the tide: recognizing climate change refugees in international law. *Law & Policy* 30(4): 502–529.

Winkler, D.W., Dunn, P.O., McCulloch, C.E. 2002. Predicting the effects of climate change on avian life-history traits. *Proceedings of the National Academy of Sciences of the United States of America* 99: 13595–13599.

WOCAT. 2007. *Where the Land is Greener: Case studies and analysis of soil and water conservation initiatives worldwide.* Liniger, H., Critchley, W. (eds). CTA: Wageningen.

Wu, H., Guo, Z., Gao, Q., Peng, C. 2009. Distribution of soil inorganic carbon storage and its changes due to agricultural land use activity in China. *Agriculture Ecosystems & Environment* 129: 413–421.

Zhang, W., Parker, K.M., Luo, Y., Wan, S., Wallace, L.L., Hu, S. 2005. Soil microbial responses to experimental warming and clipping in a tallgrass prairie. *Global Change Biology* 11: 266–277.

# 6

# RESPONSES

Given the likely sensitivity of ecosystems and human populations to interactions between climate change and land degradation, as reviewed in the previous chapter, it is essential to devise ways of mitigating these effects whilst contributing towards achieving a land degradation neutral world. Some level of adaptation[1] will also be necessary, in response to current impacts and continued likely changes arising from the effects of future GHG emissions and increases in global population (IPCC, 2014). The MA (2005) defines responses generically as "human actions, including policies, strategies, and interventions, designed to respond to specific issues, needs, opportunities, or problems". It is important to view responses in the context of perceived needs relating to the maintenance of ecosystems and populations exposed to DLDD, and the improvement of human well-being.

Climate change mitigation typically involves "an anthropogenic intervention to reduce the sources or enhance the sinks of greenhouse gases" (IPCC, 2001). To achieve LDN, mitigation of land degradation typically involves prevention of land degradation, restoration, or rehabilitation of partly degraded land and reclamation of severely degraded land, e.g. via reforestation or remediation of soils damaged by processes such as erosion or salinization (Aronson and Alexander, 2013; Grainger, 2014). Many of these mitigation responses also enhance the capacity of ecosystems and people to adapt to the dual effects of climate change and land degradation. As such, there is typically a blurring between land based mitigation and adaptation options that tackle both processes. Following the conceptual frameworks described in Chapter 3, this chapter focuses on options that can increase the adaptive capacity (and hence resilience) of ecosystems and populations experiencing climate change and land degradation.

## 6.1 Approaches to adaptation

In order for people to adapt to climate change and land degradation, they first need to perceive that something is changing, second, assess their options in light of their capabilities (the resources they have available to adapt) and third, mobilize their latent adaptive capacity to enact their adaptation decisions. Successful adaptations may be viewed as those actions that decrease vulnerability and increase resilience overall, in response to a range of immediate needs, risks and aspirations (van Aalst *et al.*, 2008), and which do not lock people into particular pathways or trajectories, or erode their future adaptive capacity.

Adaptation may be autonomous (ongoing, incremental changes to existing systems with current knowledge and technologies to cope with pressures arising from climate change and land degradation), reactive (to climate change and land degradation as they occur) or planned/anticipatory (proactive adaptations that can either adjust or transform systems at broader scales in advance of anticipated changes in climate change or land degradation) (Schneider *et al.*, 2000; Smith and Lenart, 1996; Tol *et al.*, 1998; Howden *et al.*, 2010). Due to the focus on learning from past and current adaptations to climate variability and extremes, most adaptation work has tended to focus on reactive adaptation, rather than anticipatory, planned or pro active adaptation (Reed *et al.*, 2013a). However, the use of planned strategies has been shown to enhance adaptation in many contexts (IPCC, 2007).

Adaptation can occur at a range of scales, from field-scale (e.g. changes in cropping or livestock systems, through the implementation of agroecology and climate smart agriculture approaches to cropping systems) to the implementation of policy at national and international scales (e.g. National Adaptation Programmes of Action under the UNFCCC and National Action Programmes under the UNCCD). The value of autonomous adaptations at local scales based on locally held knowledge, is increasingly being recognized, but there is recognition that effective adaptation will also require planned changes in institutional arrangements and policies to create an enabling environment for future adaptation at broader spatial scales (Stringer *et al.*, 2009; IPCC, 2014).

Béné *et al.* (2012) identify three types of adaptation: coping, adjustment and transformation. Coping is a short-term, reactive response to reduce immediate risks from climate change and drought to livelihoods. Adjustment is more commonly a planned response to longer-term climatic change and land degradation processes, and may require changes to rules, processes, structures and institutions that enable the (livelihood) system to continue functioning (Stringer *et al.*, in press). Transformation involves more fundamental changes to the social-ecological system of the governance arrangements that mediate change in the system (Klein *et al.*, 2014).

Adaptation strategies and sustainable management of agro-ecosystems cannot be developed without looking at the biophysical and sociocultural interactions and interdependencies between different ecosystems. Therefore, there is a need for a holistic approach considering both agriculture and (semi) natural (surrounding) ecosystems and their positive and negative interactions. Even if there are no apparent biophysical interactions, sociocultural factors and land use in one ecosystem may determine the outcomes of sustainable land management in another ecosystem. In that sense, not only provisioning services but also regulating, supporting and cultural services need to be taken into consideration. Only by looking at the system as a whole can we identify possible trade-offs and design most effective, efficient and sustainable adaptation measures.

Adaptation has biophysical and environmental, social, institutional, and knowledge exchange and resource needs (Burton *et al.*, 2006; IPCC, 2014):

- *Biophysical and environmental*: there is a need to enable ecosystems to adapt to the pressures of climate change and land degradation, so they can continue to provide essential ecosystem services, such as freshwater, food and climate regulation.
- *Social*: social needs for adaptation to the effects of climate change and land degradation vary geographically, with characteristics such as gender and age, and with socio-economic status. Poverty and persistent inequality underpin vulnerability to both processes, particularly for rural populations that are dependent on natural resources.
- *Institutional*: the formal and informal institutions that constrain and shape social behaviour and the institutional rules that affect negotiation and the performance of power need also to adapt (Pelling *et al.*, 2008; McGuire and Sperling, 2008). Such institutional adaptation has the potential to facilitate cross-scale solutions to climate change and land degradation, establish incentives and in other ways promote adaptation, as well as establish protocols for making and acting on decisions to adapt to climate change and land degradation (Adger *et al.*, 2005; Thomalla *et al.*, 2006; Compston, 2010). For instance, the creation and implementation of Payments for Ecosystem Services policies can be an effective stimulus for behavior change (Lapeyre and Pirard, 2013).
- *Knowledge exchange and access to resources*: successful adaptation depends on availability of and access to information, and access to technology and financial resources, from micro-finance to international financial mechanisms to facilitate adaptation to climate change and land degradation (Yohe and Tol, 2001; Adger, 2006; Eakin and Lemos, 2006; Smit and Wandel, 2006; World Bank, 2010). As part of this, the private sector may enable adaptation through Payments for Ecosystem Services schemes that deliver benefits to businesses while paying for adaptations such as agroforestry or soil management techniques that deliver climate regulation or other ecosystem services (Reed *et al.*, 2015). There is a need to enhance knowledge exchange about adaptation options for climate change and land degradation, for example combining systems such as the World Overview of Conservation Approaches and Technologies (WOCAT), and the UNCCD's various knowledge management mechanisms with systems related to climate adaptation.[2] Such efforts can be furthered by exploiting the diffusion of technologies such as mobile phones in order to raise awareness, monitor and alert land managers to changes in climate and land degradation, and increase social capital by enabling knowledge sharing about adaptations to climate change and land degradation.

In many cases, these adaptations may be informed (or triggered) by assessments of land degradation risks or climate projections (McKeon *et al.*, 2009). In some cases adaptations may harness benefits from climate change, for example by adapting the timing of crop cultivation so as to exploit longer growing seasons (IPCC, 2014). There are, however, a range of barriers to adaptation which could be addressed by changes to infrastructure, markets, access to credit, better animal health services and through the development of more effective systems for knowledge sharing (Howden *et al.*, 2010; Kabubo-Mariara, 2009; Mertz *et al.*, 2009; Silvestri *et al.*, 2012; IPCC, 2014).

It is also important to recognize the potential for maladaptation. This is explored later in this chapter. One way of avoiding maladaptation is to identify "no-regret", "low regret" or "win win" adaptation options. "No-regret" options should be implemented irrespective of climate change and land degradation, and are considered to be politically and economically feasible under a range of possible future climates. "Low regret" options are cost-effective and low-risk responses to climate change and land degradation with relatively large benefits for vulnerable sectors, geographical regions or populations. "Win-win" options are those that contribute to adaptation but also have wider social, environmental or economic policy benefits, including mitigation benefits (de Bruin *et al.*, 2009; Stringer *et al.*, in press). The final section of this chapter considers whether it may be possible to develop "triple-win" options that provide opportunities for both mitigation of and adaptation to climate change and land degradation.

While climate change is currently the global focus, neglecting natural climate variability in land management may sometimes be a precursor to land degradation, which is then exacerbated when long-term climate change occurs. Conversely, if land managers are prepared for short-term climate variability, something they themselves can do with improved land management, they are better prepared for long-term climate change. However, identifying ways to enhance implementation of climate change adaptation in order to help tackle desertification at the ground level remains a challenge (Seely and Montgomery, 2011).

## 6.2 Options for simultaneously adapting to climate change and land degradation

A range of adaptation options have the potential to tackle land degradation and climate change simultaneously, and may therefore be able to mitigate some of the interactions between these two processes and protect livelihoods and human well-being. These include:

- adaptation of cropping systems;
- adaptation of livestock systems;
- climate-smart agriculture;
- ecosystem-based adaptation; and
- sustainable land management.

Each of these will now be discussed in turn.

### 6.2.1 *Adaptation of cropping systems*

There is already evidence that farmers are adapting to climate change by changing cultivation and sowing times, crop cultivars and species, and developing new marketing arrangements (Fujisawa and Koyabashi; 2010; Olesen *et al.*, 2011; IPCC, 2014). Many SLM options are also already being used to adapt to the effects of land degradation. For example, to adapt to soil degradation people are using agroecology and agroforestry techniques, such as intercropping with leguminous woody species. This allows access to nutrients deeper in the soil profile, whilst simultaneously reducing the effects of erosion and increasing levels of soil fertility higher up the soil profile. Another adaptation is the on-farm production and use of organic materials (compost, vermicompost, biochar and other by-products) which can improve soil fertility. The use of different types of biofertilizers can increase plant nutrient mobilization, nitrogen fixation and demineralization, while adaptation of the tillage system (including no-till and conservation tillage) can improve the soil quality. The benefits of such practices nevertheless have to be carefully evaluated in relation to labour costs, as well ensuring there are no detrimental effects in terms of moisture or light availability.

### 6.2.2 *Adaptation of livestock systems*

In many cases, livestock systems are already highly adapted to climate variability, and provide a valuable means of adapting to future climate change, often drawing on locally held knowledge to inform decision making. A range of adaptations to livestock systems can enable adaptation to both land degradation and climate change (IPCC, 2014), including:

- altering stocking rates to match changes in forage production in response to climate change and/or land degradation;
- adjusting the management of herds and water points in response to changing seasonal and spatial patterns of forage production under climate change and inter-annual trends in forage production due to land degradation;
- managing diet quality (using dietary supplements, legumes, choice of introduced pasture species and pasture fertility management) to maintain herds under climate change and/or land degradation;
- more effective use of rotational grazing systems;
- managing the encroachment of woody shrubs spreading on productive rangeland;
- using livestock breeds or species that are better suited to new conditions as a result of climate change and/or land degradation;
- increased provision of shade from trees to reduce heat stress in livestock though the adoption of silvopastoral systems that can also reduce erosion rates and provide fodder for livestock during drought;
- enabling migratory pastoralist activities (though this has to be carefully managed to avoid exacerbating land-use conflicts);

- monitoring and managing the spread of livestock and rangeland pests, weeds and diseases; and
- improved soil and water management.

### 6.2.3 Climate-smart agriculture

Adaptation of cropping and livestock systems should be considered within the emerging concept of Climate-Smart Agriculture (CSA), as defined and presented by FAO at the Hague Conference on Agriculture, Food Security and Climate Change in 2010. It integrates the three dimensions of sustainable development (economic, social and environmental) by jointly addressing food security and climate challenges. It is composed of three main pillars:

- sustainably increasing agricultural productivity and incomes;
- adapting and building resilience to climate change; and
- reducing and/or removing greenhouse gases emissions, where possible.

Where climate change and/or land degradation threatens current livelihood strategies, CSA may include diversification into new sources of income to increase the resilience of the system and further support livelihoods. CSA is an approach to developing the technical, policy and investment conditions to achieve sustainable agricultural development for food security under climate change. The magnitude, immediacy and broad scope of the effects of climate change on agricultural systems create a compelling need to ensure comprehensive integration of these effects into national agricultural planning, investments and programmes. The CSA approach is designed to identify and operationalize sustainable agricultural development within the explicit parameters of climate change.

### 6.2.4 Ecosystem-based adaptations

Some adaptations that use biodiversity and ecosystem services to enable adaptation to climate change may also be able to enable adaptation to the effects of land degradation. For example, wetland restoration may be able to provide water resources for livestock and cropping systems, whilst creating a buffer to climate-induced flood risks (Huntjens et al., 2010; Jones et al., 2012). Green infrastructure, such as green roofs, porous pavements and urban wildlife corridors, can reduce soil-sealing whilst improving storm water management, reducing flood risk in cities, and moderating the heat-island effect (IPCC, 2014). Ecosystem-based adaptations such as these have the potential to simultaneously enable adaptation to climate change and land degradation, whilst in many cases also protecting and enhancing biodiversity.

### 6.2.5 Sustainable land management

A range of adaptations to soil and water management practices can enhance adaptation to both climate change and land degradation. These include, for example, building terraces or other structures that can reduce erosion and tackle land degradation whilst also mitigating downstream flood risk as a result of changes in rainfall patterns under climate change. These options are considered in more detail in the rest of this section.

It is now widely acknowledged that Sustainable Land Management[3] (SLM) can simultaneously tackle land degradation, reduce net GHG emissions and contribute towards the conservation of biodiversity, thereby contributing towards the goals of all three Rio Conventions (Thomas, 2008; Cowie *et al.*, 2011) and SDG target 15.3 on LDN. Whether these "wins" can be achieved whilst still protecting food production, livelihoods, social equity, economic viability and cultural values depends on the ways in which they are enacted, and requires a delicate balance to be reached. A range of SLM technologies[4] can be used within an overall SLM approach. These technologies need to be viewed in the context of their socio-cultural and policy environment, which may enable or hinder their development and adoption. Rather than attempting to provide an overview of SLM, this section focuses on how SLM might be able to mitigate interactions between climate change and land degradation.

SLM technologies typically attempt to maintain a protective biological surface cover (e.g. living plants or mulches), good soil structure and adequate levels of soil organic matter (Thomas, 2008). As a result, such measures also typically reduce GHG emissions from agriculture. Thomas (2008: 597) argues that

> maintaining a cover over the ground and developing a better stewardship of the flora and fauna will help prevent and reverse land degradation, increase the resilience of ecosystems to climatic and human-induced stresses and thereby contribute to the conservation of biodiversity and mitigation of climate change.

Increasing soil carbon enhances infiltration and moisture retention, and therefore may improve water availability for crops and forage during drought (Cowie *et al.*, 2011).

In this way, SLM can help avoid the feedback between climate change and land degradation via changes in vegetation and soil carbon stocks. Rather than losing carbon to the atmosphere due to land degradation, and contributing towards climate change, SLM can build resilience in the face of climate change by increasing soil organic matter (Aguilera *et al.*, 2013). It has further been argued that SLM at a global scale has the potential to sequester and store significant amounts of carbon, thereby helping to mitigate climate change (Lal, 2004; 2007). SLM practices can also directly link to the feedback between climate change and land degradation that is mediated through losses of vegetation cover. Rather than losing biomass and vegetation cover, which can lead to regional climatic changes (including drier conditions at particular times of year), SLM maintains biomass and vegetation cover, and so contributes towards more stable regional climates. Although SLM would of course reduce the albedo-based cooling effect of land degradation at a global scale, this would be offset to an unknown extent by its carbon sequestration benefits. Finally, certain SLM technologies also have the potential to mitigate biodiversity-mediated feedbacks between climate change and land degradation. For example, cover crops and mulches can have benefits for plant diversity and create habitats for arthropods (Andersen *et al.*, 2013; Bryant *et al.*, 2013; Scopel *et al.*, 2013; Licznar-Małańczuk, 2014).

SLM technologies often evolve through local traditional practices and incremental experimentation rather than being taken up on the basis of scientific evidence (Berkes *et al.*, 2000). SLM technologies are also often suited to particular biophysical or socio-cultural contexts (Liniger and Critchley, 2007). These factors make it difficult to effectively promote the adoption of these technologies and scale up SLM from field to regional and national scales (Stringer and Reed, 2007). Other barriers to the adoption of SLM technologies include for example, the cost of introducing or maintaining the technology; availability of labour to implement it; local traditions and cultural factors; or logistical challenges such as distance to markets (Stringer *et al.*, 2014). These complications make it imperative to combine locally held knowledge on SLM technologies with scientific testing and validation, so that local technologies and know-how can gain greater policy credence and be more widely applicable across contexts (Raymond *el al.*, 2010; Stringer *et al.*, 2014). Despite the challenges associated with achieving SLM, there is already a wide range of technologies available (see Box 6.1), alongside growing awareness of the constraints that are preventing more widespread uptake. These will be considered later in the chapter.

## BOX 6.1: A DESIRE FOR GREENER LAND – AN INTERNATIONAL ASSESSMENT OF LAND DEGRADATION AND SUSTAINABLE LAND MANAGEMENT OPTIONS

The EU-funded DESIRE project (2007–2012) developed an approach to establish promising SLM strategies in response to land degradation in drylands. The DESIRE approach consists of five steps:

(1) establishing land degradation and SLM context and sustainability goals;
(2) identifying, evaluating and selecting SLM strategies;
(3) trialling and monitoring SLM strategies;
(4) upscaling SLM strategies; and
(5) disseminating the knowledge gathered in the previous steps.

The DESIRE approach was applied in 17 areas affected by desertification, accounting for a wide variety of biophysical and socioeconomic conditions found worldwide. The DESIRE approach can be applied by agricultural advisors, government institutions, or in any project that aims to combat land degradation. To date, it has been incorporated in publications and initiatives by the UNCCD, FAO and the Global Environment Facility (GEF).

During the project, mapping of land degradation and current SLM showed that land degradation in the 17 DESIRE study sites mainly occurred as water erosion on cultivated land and land under mixed use. Degradation was found to be increasing in most sites, primarily caused by inappropriate soil management. Indirectly, land degradation appeared to be caused most frequently by population pressure, insecure land tenure and poverty, in combination with aspects of governance, institutional functioning and politics. The SLM measures already found to exist in the study sites mainly comprised grazing land management technologies and conservation agriculture. Combinations of SLM measures appeared to perform better than applying one type of measure by itself.

Land degradation negatively affected ecosystem services for almost all degraded areas. Provision of ecosystem services – such as food, fodder, wood, water and energy – was most affected in areas of mixed land use, followed by areas of cultivated land and grazing land. High negative impacts were observed regarding regulation of ecosystem services – such as regulation of water and nutrient flows, carbon sequestration, pollination and pest control – indicating that these require particular attention when developing and implementing remediation strategies. SLM measures appeared most effective on cultivated land, but positive impacts from SLM on ecosystem services were also recorded for relatively large areas of forest and grazing land. Overall, there appears to be scope for improving SLM contributions to ecosystem services in cultivated land.

There were 38 case studies investigated in the DESIRE project. The physical practices used in the field to control land degradation and enhance productivity – the SLM technologies, in other words – could be divided into five groups: cropping management, water management, cross-slope barriers, grazing land management and forest management. They addressed all the main types of land degradation. Most of them were applied on cropland, although grazing land is equally important – perhaps even more important in spatial terms – in drylands. Depending on the kind of degradation addressed, agronomic, vegetative, structural or management measures were used, or some combination of these. Most of the technologies aimed to prevent or mitigate degradation; only a few were described as intended for rehabilitation. This reflected the state of land degradation in the study sites, which had not passed thresholds of extreme loss of productivity or provision of ecosystem services. The main functions of the SLM technologies assessed included their ability to increase infiltration capacity, control runoff and improve ground cover. Most of the technologies were applied by small-scale land users, a group that is often underestimated regarding investment and innovation, not to mention their role in worldwide agricultural production.

The SLM technologies positively affected biophysical processes relevant for agricultural production and positively affected the ecological services of the land.

Water harvesting technologies and more efficient use of irrigated water showed the greatest potential and benefits. Most of the applied technologies appear resilient to expected climatic variations and half of them provide off-site benefits, such as reduced damage to neighbouring fields, public or private infrastructure, and reduced downstream flooding.

Adapted from: Schwilch, G., Hessel, R. and Verzandvoort, S. (eds). 2012. *Desire for Greener Land. Options for Sustainable Land Management in Drylands.* Bern, Switzerland, and Wageningen, The Netherlands: University of Bern – CDE, Alterra – Wageningen UR, ISRIC – World Soil Information and CTA – Technical Centre for Agricultural and Rural Cooperation.

## 6.3 Learning to adapt using locally held and scientific knowledge

There is a danger that adaptations based on scientific knowledge alone may not be suitable for the socio-cultural context in which they are needed, and this may significantly limit their uptake and effectiveness. By combining scientific understanding of adaptation options with local, contextual knowledge, it may be possible to develop adaptations based on generations' worth of experiential knowledge, which can help refine adaptations.

Local knowledge is often used synonymously with traditional knowledge (sometimes called indigenous knowledge), yet a subtle distinction can be made. Both are highly context-specific, but local knowledge emphasizes the geographical provenance of the knowledge, whereas traditional knowledge emphasizes historical and cultural provenance (Raymond *et al.*, 2010). For example, Olsson and Folke (2001) suggested that a local fishing association in a Swedish community displayed management practices that enabled the protection of crayfish beyond the local crayfish population to the wider ecosystem, so this provides an example of local ecological knowledge. This may be contrasted with traditional knowledge. For example, the Turkwel Riverine Forest in Kenya has been managed for many years by an indigenous system known as *ekwar* which refers to a parcel of riverine forests whereby the owner and family has exclusive rights to collect building materials, firewood and edible fruits. Outsiders require permission from the *ekwar* owner to graze their livestock in the area (Stave *et al.*, 2007).

Local and traditional knowledge can provide a wealth of adaptation options based on previous experience and exposure to climate change and land degradation processes (Dixon *et al.*, 2014). For example, traditional methods of water harvesting, making use of local topographic and soil characteristics, or using architectural innovations to condense atmospheric water (e.g. stone heaps, dry walls, little cavities and depressions in the soil) can successfully enable plants to overcome periods of drought and improve productivity (Biazin *et al.*, 2012; Box 6.2). Inca traditions of crop diversification, raised bed cultivation, agroforestry, weather forecasting and

## BOX 6.2: WATER HARVESTING

Water security is an important first step towards achieving food security and is under threat in many parts of the world from predicted climate change. Rain-fed agricultural systems are particularly vulnerable. Over 80 per cent of global agriculture depends on rainfall alone, contributing to at least two-thirds global food production (FAOSTAT, 2005, as cited in Rockström *et al.*, 2007). In developing countries, including in Latin America and sub-Saharan Africa, over 90 per cent of agriculture is rain-fed. Where climate-induced water shortages combine with ongoing land degradation processes in these systems, agricultural productivity and livelihoods may be threatened. Irrigation water is only available to a minor proportion of smallholders and has been associated in the past with over-abstraction, soil salinization and high costs. Water harvesting on the other hand, does not rely on fossil water, is less likely to lead to salinization and is more likely to be affordable. It can help tackle land degradation by capturing water from otherwise highly erosive heavy rainfall events and storing the water to help buffer livestock and crops against drought. In this way, water harvesting helps buffer agricultural systems from extreme weather events that are predicted to become more prevalent under climate change.

Many different water harvesting techniques have been developed by local communities around the world over many centuries to adapt to climatic variability. Such techniques are increasingly being recognized for their potential to enable communities to adapt to longer-term climate change. The World Overview of Conservation Approaches and Technologies (WOCAT) identifies four types of water harvesting technologies (Mekdaschi Studer and Liniger, 2013):

- flood water harvesting (e.g. spate irrigation, floodwater farming and water-spreading weirs);
- macro-catchment water harvesting (e.g. sunken streambed structures, small earth dams, sand dams, recharge wells);
- micro-catchment water harvesting (e.g. planting pits, furrow enhanced runoff harvesting, the Vallerani system); and
- rooftop and courtyard water harvesting (e.g. various designs of rooftop rainwater harvesting systems).

Water harvesting technologies typically consist of: (i) a catchment or collection area (e.g. a rooftop or rangeland); (ii) a conveyance system (e.g. gutters, pipes or furrows); (iii) a storage component (e.g. jars, tanks, ponds and reservoirs); and (iv) an application or target area (e.g. for domestic or livestock consumption or for irrigation of arable crops) (Mekdaschi Studer and Liniger, 2013).

water harvesting are still used today in the southern Andes (Goodman-Elgar, 2008; McDowell and Hess, 2012).

Locally held (including traditional) knowledge is typically context-specific, and may have limited potential for use elsewhere, but it is often highly relevant and acceptable within the socio-political and biophysical context in which it was developed (Raymond *et al.*, 2010). This kind of knowledge is typically highly dynamic, and involves learning from local experimentation and incorporating ideas from other areas (for example observed during seasonal or temporary migrations) (Mazzucato and Niemeijer, 2000). It has informed responses to past climate variability and land degradation, leading to claims that these knowledge systems can help to foster learning and livelihood resilience in the face of future climate and land degradation risks (Stringer *et al.*, 2009). However, little work has been done to assess the extent to which locally held knowledge can inform future adaptations. In particular, there are concerns that the future may see extremes that exceed those that have been experienced in the past, reducing the utility of local and traditional knowledge (Speranza *et al.*, 2010; Kalanda-Joshua *et al.*, 2011; McDowell and Hess, 2012).

These observations support the need for scientific approaches to play a key role in generating new adaptations to climate change and land degradation, using some of the methods and modelling techniques mentioned in the methodological framework described in Chapter 3. Science can provide certain kinds of information that can be difficult to capture through locally held knowledge alone, for example, providing data at spatial and temporal scales that would otherwise not be possible to consider. By elucidating the processes through which climate change and land degradation impact upon livelihoods, scientific evidence can identify potential system feedbacks and unanticipated impacts that can inform the development of adaptation options (Reed *et al.*, 2011).

Although adaptations based on scientific knowledge are typically more widely applicable with greater potential for use in different contexts, they may have limited social and cultural acceptance. It is therefore essential to link different knowledge systems to foster adaptation, putting scientific findings alongside local knowledge, recognizing each as valid (whether formally codified or not), and scrutinizing each with equal rigour (Berkes and Folke, 2002; Raymond *et al.*, 2010; Stringer *et al.*, 2014). Thomas and Twyman (2004) combined local and scientific knowledge to generate what they called "hybrid" knowledge, to interpret whether changes in rangeland in the Kalahari desert could be considered "good" or "bad" (i.e. degraded or not). Following this approach, scientific evidence may be questioned and perhaps rejected in favor of adaptations based on local experience that are more profitable, less risky or more culturally acceptable. Such context sensitivity is vital in instances where behavioural changes are needed. In cases where local approaches are perceived to be failing, it is important to distinguish cases where the technologies work but need a more enabling political or socio-economic environment, from those cases where more appropriate technologies may need to be developed (MA, 2005).

Raymond *et al.* (2010) nevertheless argue that categorizing knowledge as "local", "scientific" or "hybrid" is too simplistic, as it does not sufficiently take into account the way individuals learn, make sense of new information, or the social contexts that influence how people understand something. They therefore propose that when attempting to integrate different types of knowledge, time should be spent identifying, characterizing and evaluating the different knowledges involved and how they might be relevant. For example, this might involve ensuring that experts engaging in the process have sufficient depth of experience directly relevant to the problem to be addressed (Fazey *et al.*, 2006). It may also be necessary to ensure that the range of stakeholders involved in the elicitation and integration of local and traditional knowledge can provide relevant knowledge and expertise, such as ecological or economic expertise, that can help to improve understanding of the inter-related human and social aspects of a system or problem.

A number of studies have attempted to compare, contrast and combine different types of knowledge about land degradation and climate change. For example, Oba and Kotile (2001) compared assessments of land degradation by ecologists with methods based on the local knowledge of Booran pastoralists in Ethiopia. The results of assessments by local pastoralists strongly correlated with results from range ecologists, with comparable assessments of rangeland condition and trends arising from each group. However, each group interpreted these changes differently, with range ecologists more likely to identify stable ecosystem states in terms of their temporal variability rather than their utility for livelihoods. This meant that a stable bush vegetation state at "ecological climax" would be considered "stable" by the ecologist but "deteriorating" by the pastoralist. Similarly, Forsyth (1996) used a combination of local and scientific knowledge to challenge a politically motivated discourse about the degradation of Himalayan farmed uplands, showing that local communities were practicing conservation agriculture and supporting local claims with scientific evidence using the Caesium-137 technique to show that downstream sediments did not originate from upland farmlands.

## 6.4 Overcoming barriers to adaptation

It is important to note that there may be limits to adaptation, and that the extent and speed of climate change and land degradation in some locations may be unlike anything ever experienced previously. This may limit the capacity to base future adaptive strategies on lessons from past experience. For example, using model-based approaches, Thomas *et al.* (2005) suggested that a loss of biomass to below 14 per cent vegetation cover combined with a temperature increase of 2.5–3.5°C would lead to dune reactivation across much of the Kalahari Desert in southern Africa by 2100. This future scenario would likely be hastened by continued grazing by livestock, as herds are maintained by groundwater and supplementary feeding during increasingly frequent and extended droughts. With insufficient forage available to support livestock, adaptation options would be increasingly constrained.

There may also be barriers to adaptation, e.g. limited land area and inputs with which to increase agricultural production, limited human capital in terms of labour or time, or limited financial capital to invest in diversification options (Suckall *et al.*, 2014). The extent to which individuals, households and communities can adapt their livelihoods in response to climate change and land degradation, therefore may be restricted in a number of different ways. Furthermore, not all adaptations are successful and some can increase vulnerability over the longer term. It is therefore important to take a critical approach to the assessment of adaptation, considering also the potential for maladaptation, and possible limits and barriers with regard to what adaptation is able to achieve (Stringer *et al.*, in press).

Barriers to adaptation can arise from a variety of sources, for example: a lack of available options to substitute one form of capital for another (e.g. due to a limited asset base, limited agro-ecosystem capacity or limited market access); limited political capacity to enact strategies to support adaptation; high levels of institutional inertia and rigidity; lack of access to information about adaptation options (including poor agricultural extension services); or financial constraints (including lack of access to credit) (Deressa *et al.*, 2009; Kabubo-Mariara, 2009; de Bruin and Dellink, 2011; Quinn *et al.*, 2011; Silvestri *et al.* 2012). Other barriers can be cognitive in nature, linked to a lack of perceived risk, a lack of perceived agency and a sense of powerlessness, or the social norms that influence behaviour within particular socio-cultural settings (MA, 2005), or a lack of limited incentives to change behaviour. For example, opportunities for women to diversify into new livelihood activities could be restricted due to particular cultural or religious expectations (Stringer *et al.*, in press).

Important barriers to the adoption of adaptation measures include a lack of awareness, lack of available knowledge, and differences in perception of problems and solutions to the impacts from climate change and land degradation by different stakeholders. That land degradation and climate change are generally slow, creeping and complex processes often hampers fast adoption of adaptation measures by stakeholders who may perceive other priorities as more urgent.

However, there is increasing empirical evidence that well-designed participatory processes may help to overcome some of these barriers to adoption, as participatory processes often lead to social learning, increased trust between stakeholders and ownership over problems and solutions. This means that decisions taken in participatory processes are more likely to be accepted and implemented (Reed, 2008; de Vente *et al.*, in press). As an example, the Regional Landcare Facilitator (RFL) programme is an initiative of the Department of Agriculture of Australia. This 4-year programme funds Regional Landcare Facilitators located in each of the 56 natural resource management regions across Australia where they support Landcare (a unique grass-roots movement that started in the 1980s through initiatives to tackle degradation of farmland, public land and waterways) and production groups to adopt SLM practices and to protect Australia's landscape. The Natural Resource Management PlaceStories Project provides members with a powerful digital storytelling and communication digital platform to help natural resource managers to document and report on their natural resource management work and projects, monitor and evaluate management activities; share successes and key learnings with others, promote effective practices, and communicate and collaborate online (see http://placestories.com/project/8169). Such knowledge exchange activities can provide an excellent starting point for the development of partnerships, as stakeholders move towards shared goals (Box 6.3).

Other barriers and limits are associated with a range of adaptation options and contexts. Broadly speaking, adaptation options work within the constraints of the capital assets available to the individual, household or community. So for example,

---

## BOX 6.3: PARTNERSHIPS FOR COMMON GROUND

Many barriers to adaptation are linked to particular resource, regulatory, participatory and learning gaps (Pinske and Kolk 2012). Partnerships between different stakeholders can often allow the strengths of each group to be harnessed, in order to deliver outcomes that would not be otherwise possible. For example, Dyer et al. (2013) examined the Kansanshi Foundation Conservation Farming Project in Zambia, which was established in 2010. This initiative was primarily funded by a mining company as part of its corporate social responsibility activities, but was undertaken in partnership with the Zambian government's Ministry of Agriculture and Livestock, the traditional authorities and the communities in the target area. The initiative aimed to diversify livelihood activities in communities close to the mine, whilst also reducing deforestation driven by charcoal production (making a positive contribution to climate change mitigation), and enhancing adaptation options by improving food production and upscaling the use of soil conservation techniques such as reduced tillage, crop rotation and cover crops. Supporting infrastructure included training on conservation farming, as well as a micro-credit scheme for fertilizer and maize seeds. Within the partnership, the mine provided finances, equipment and expertise for conservation farming (addressing resource and learning gaps); the Ministry assisted with capacity building (addressing resource and learning gaps) and the implementation of the conservation farming project through the use of government agricultural extension networks to gain access to communities (addressing a participatory gap). Communities provided local knowledge and labour (targeting resource gaps) as well as granting the mine a social licence to operate (addressing a regulatory gap). The traditional authorities gave legitimacy to the project and encouraged communities to participate (addressing participatory and regulatory gaps). Although not without its challenges (see Dyer et al., 2013), the project provides a useful example of how partnership working can be used to overcome barriers to adaptation and create a shared focus for multi-stakeholder working.

agricultural intensification may be a relevant adaptation to climate change and land degradation in some areas, but will be limited by natural capital in terms of the soil nutrients or water that is available. However, it may be possible to draw upon human capital to develop rainwater harvesting techniques to overcome water shortages or to use financial capital to purchase fertilizer. In other areas where population densities are low, extensification may be considered as an adaptation option. However, this may be constrained by human capital if there is insufficient labour to herd animals (Stringer et al., in press). Where adaptations depend on the use or substitution of capital assets, it is necessary to unravel how this asset base is likely to be affected by climate change and land degradation, in order to understand future adaptive capacity (Reed et al., 2013a). It is also important to assess how climate change and land degradation are likely to influence people's abilities to access or substitute assets if maladaptation is to be avoided.

Maladaptation may for example: increase GHG emissions (e.g. via fossil fuel use by desalinization plants); increase polarization between rich and poor or disproportionately burden the poor (e.g. by raising the costs of water and energy or privatizing communal rangeland); lead to high opportunity costs (whether economic, environmental or social costs); and create path dependencies, where communities are locked in to particular technologies or livelihood strategies that may compromise their capacity or willingness to adapt in future (Barnett and O'Neil, 2010; Pittock, 2011). Many maladaptations result from poor investment decisions and substitutions that do not remove risks and threats but merely shift them elsewhere, to other sectors, other people and other places. Similarly, a maladaptation to exposure to one threat can leave the same people or location more vulnerable to other threats. For example, farmers in Malawi who are experiencing drought and associated low yields may temporarily migrate to seek employment on sugar plantations in neighbouring Mozambique. While this might start out as a short-term coping strategy to raise funds to buy food, it removes labour from people's own land. This not only leaves them vulnerable to weed infestations that could result in low yields in subsequent years, but also means they are affected by the vagaries of market dynamics when purchasing food.

Reed et al. (2013a) characterize adaptation decisions as a choice between: (i) adopting adaptations based on new ways of using or substituting between existing assets; or (ii) developing new assets. In some cases, opportunities to develop new assets and associated livelihood options may arise as a consequence of climate change, for example cultivating new crops. Although these adaptations may have been tried elsewhere, they are innovations in a new context or environment or for a different social group. As such, it may be useful to think about the evaluation of adaptation options as an innovation-decision, in which the perceived relative advantage, trialability, compatibility, observability, complexity and adaptability of different options are evaluated against each other and current practice (Rogers, 1995). The literature on social learning and the diffusion of innovations emphasizes that such decisions are evaluated in a social context. For example, people's decisions are influenced by others to whom they are socially tied. Social networks, in other words, influence how individuals learn and consequently make decisions (Prell et al., 2009). This happens for a number of reasons: social psychology and social network analysis research shows that individuals tend to adapt their views to those around them as a way to decrease cognitive dissonance (Homans, 1950; Friedkin, 1998; Ruef et al., 2003; Skvoretz et al., 2004).

Adaptation may also be constrained by institutional and structural barriers, for example linked to land tenure, or globalized processes of 'land grabbing' which limit access to the natural resource base (i.e. large-scale acquisition of arable land by foreign companies or governments, typically for cash crops) (Stringer et al., in press). For example, the "Land Policy Initiative" (www.uneca/lpi) aims to ensure that all land users have equitable access to land and security of land rights to boost sustainable development in Africa. National policies can both incentivize and act as barriers to adaptation, for example driving the production of certain crops, and so simplifying the agro-ecosystem and limiting future adaptation options (Dixon et al., 2014). Such policy decisions can create path dependencies and lock-in effects, which further limit future adaptation (Bailey and Wilson, 2009; Wilson, 2014). For example, Freier et al. (2012) describe how despite predicted decreases in rainfall, pastoralists in semi-arid Morocco still chose livelihoods based on livestock, and were more likely to abandon nomadic lifestyles, even though this would increase pressure on available land and water resources, and reduce resilience. Equally, informal institutional barriers may limit the range of adaptive options that are considered to be acceptable, if adaptation options are not compatible with prevailing social norms and customs. As such, many adaptations require a certain degree of behavioural change.

Adaptations that require behavioural change within individuals, households, communities or institutions are often linked to transformative adaptation (Stringer et al., in press). Behavioural adaptations often require changes from a previous or current activity or way of doing an activity to a new one. Short-term behavioural changes can sometimes be top-down, and shaped by prevailing institutions and laws, by, for example, restricting access to certain areas to allow ecosystem recovery following drought. Such changes may be effective in improving the long-term ecosystem state and allowing vegetation recovery, but can have inequitable and unjust implications for some groups over the short and medium term – particularly those whose livelihoods depend solely on access to those areas.

Long-term behavioural changes can include adopting new agricultural techniques, switching to growing different crops, or changing planting and/or harvesting calendars. However, these kinds of household level changes often also require wider institutional support. For example, while a household may recognize the need to plant earlier and mobilize the financial capital necessary to hire a tractor to prepare the soil earlier than is usually the case, unless the institutions responsible for making tractors available have sufficient accessible machinery to meet changing timings of demand, the household will be unable to enact their adaptation decision (Simelton et al., 2013). This kind of scenario is particularly problematic when several households decide to undertake the same adaptation at the same time, placing unexpected new demands on particular institutions. Such relations between adaptation practice and these broader institutional factors demonstrate the importance of appropriate policy instruments and governance models in enacting adaptive capacities.

Reed *et al.* (2013a) argue that adaptation decisions are influenced by the aspirations of the decision-maker. This views livelihood decisions as aiming for a satisfactory outcome (defined by an aspiration level) rather than necessarily the optimal outcome (a process sometimes called "satisficing"; Simon, 1955; 1956). If livelihoods are sensitive to climate change, a reduction in assets may be deemed acceptable to an actor with a low aspiration level who would perceive no need to adapt. However, the same reduction in assets may stimulate a search for adaptive options by the same actor if their aspiration levels were higher. Different adaptation options may be necessary to meet different aspiration levels. In this context, adaptation may be used to improve rather than simply maintain livelihoods in the face of future change (Ziervogel *et al.*, 2006). If livelihood outcomes are no longer deemed satisfactory, then a search commences for livelihood adaptation options, which are evaluated against individual decision rules.

Limited research to date has sought to examine the links between the adaptations taking place on the ground and the support for adaptation provided within policy. Stringer *et al.* (2009) examined local adaptations as well as policy adaptations as outlined in UNCCD NAPs and National Communications to the UNFCCC in three countries in sub-Saharan Africa. Their research found that there are some overlaps in terms of the types of adaptations in policy and practice, but that these are largely coincidental rather than the result of the active incorporation of local adaptations within policy planning. Similar to other research (e.g. Kalaba *et al.*, 2014), a need for mainstreaming adaptation within policy is recognized, in order to avoid duplication of efforts. A lack of mainstreaming also increases the risk of negative externalities and can prevent policies from undermining the success of other policies and strategies in other sectors. Policy lock-ins are also important as existing strategies are reinforced over time, such that resistance to change can develop. Policies have further been criticized for their lack of consideration of social and cultural factors. For example, in some countries, micro-credit schemes have been proposed as an important policy mechanism to address shortages of financial capital at key points in the agricultural calendar when people require the purchase of agricultural inputs. However, many rural people are disinclined to engage in such schemes, particularly where it would require them to use cattle as collateral.

To overcome these challenges, there is a need to bring together top down policy approaches across key livelihood sectors, with bottom up adaptations that are already taking place on the ground. This could help people to enact the successful adaptations that they would choose to employ, and which are already tailored to the realities of the local livelihood and social-cultural context. It is nevertheless important to recognize that not all adaptation options are compatible with one another and what is a successful adaptation at one scale could undermine adaptation options at larger scales, particularly under conditions of future uncertainty.

For example, agronomic adaptations, such as changing sowing dates, supplementary feeding of livestock and irrigation may help sustain agricultural productivity under climate change. However when interactions with land degradation processes are considered, the benefits may be short-lived, leading to far worse degradation of agricultural land in the long-term than would have been seen from the effects of climate change alone. For example, irrigation may provide short-term gains but ultimately lead to soil salinization, especially if using groundwater from coastal aquifers that are threatened by salt-water intrusion due to sea level rise. Similarly, maintaining livestock numbers through drought by supplementary feeding can undermine the natural resilience of some ecosystems, replacing perennial grasses with less palatable annuals or leading to thorny bush encroachment. Drylands are often naturally able to bounce back effectively from drought, and although there would probably be a gradual shift towards species that are suited to more arid climates, the productivity of most drylands would continue to recover after droughts under future climate change. But their ability to do this will depend on the way people manage the land during drought as well as afterwards when the phase of natural regeneration starts. Depending on our actions, climate change may or may not lead to medium- or long-term, irreversible land degradation.

Payments for Ecosystem Services (PES) is a policy instrument that can help overcome obstacles to SLM adoption and other ecosystem-friendly behaviours (Pirard et al., 2010). When designed as asset-building mechanisms (i.e. paying farmers conditionally to the adoption of new farming, or other natural resource use, techniques), PES may assist rural communities, especially the poorest and most vulnerable, with start-up physical and financial capital, as well as training, in order for them to adopt new and sustainable strategies. Without such technical and financial assistance during the start-up phase, farmers would not modify their strategies, even if this behaviour change is beneficial to them in the longer term. Such PES instruments, as already implemented currently, include PES schemes for stimulating adoption of silvopastoral conservation practices in Costa Rica, Nicaragua and Colombia (Garbach et al., 2012).

So far, the focus has largely been on complementarities between adaptation options for both climate change and land degradation. However, it is also necessary to evaluate potential trade-offs between adaptations, so that complementary bundles of adaptations can be implemented together, avoiding maladaptation and reducing vulnerability to both climate change and land degradation. A variety of techniques have been developed for systematically selecting adaptation options (e.g. de Bruin et al., 2009; Ogden and Innes, 2009; Füssel, 2009). However, each has its limitations, as they are typically unable to account for many of the barriers reviewed in this section (Smith et al., 2009). In addition, most economic techniques struggle to capture non-market costs and benefits, which can be considerable (IPCC, 2014).

Evidence from studies of adaptations to past and current climate variability and extremes show that adaptation options are rarely adopted singly (e.g. Reid and Vogel, 2006; Stringer et al., in press). Instead, bundles of complementary adaptation options are adopted together, overlapping in time and space, in an attempt to address the multiple outcomes of climate change and land degradation. However, not all adaptation options are necessarily compatible with one another, and it is important to investigate in advance the likely effects of combining different adaptations to climate change and land degradation. For example, by coupling Agent-Based Models with biophysical (e.g. soil erosion) and climate models, it is possible to project which adaptation options are likely to be adopted where (e.g. Fleskens et al., 2014). Reed et al. (2013b) developed adaptation options on the basis of combined land degradation and climate scenarios and considered likely consequences for biodiversity, and Ceccarelli et al. (2014) developed a range of land degradation (principally land abandonment and soil sealing) and climate change scenarios. However, there have been few attempts to consider how climate and land degradation adaptation options might interact in space and time. Such an analysis would facilitate the development of complementary bundles of options to reduce the vulnerability of ecosystems and populations to both climate change and land degradation, whilst considering the likely effects on biodiversity. Evidence presented in this chapter suggests that it is theoretically possible to develop adaptations to both climate change and land degradation, which can in many cases have benefits for biodiversity. Ecosystem-based approaches and SLM have particular potential in this regard. It may therefore be possible in future to develop "triple-win" options that enable adaptation to climate change, land degradation and biodiversity (Suckall et al., 2014).

## 6.5 Synthesis

This chapter has considered how adaptive capacity may be enhanced to retain the integrity of ecosystems in regions affected by DLDD and maintain sustainable livelihoods in the face of the interactive effects of climate change and land degradation. It has reviewed different approaches to adaptation, including: autonomous, reactive and planned/anticipatory adaptation; coping, adjustment and transformation; and win-win versus no-regret and low-regret adaptation options. There is a range of barriers to adaptation, and there is also the potential for maladaptation and trade-offs between adaptation options. We have assessed adaptation needs in relation to climate change and land degradation, in terms of biophysical and environmental, social, institutional, and information, capacity and resource needs.

The chapter has considered options for simultaneously adapting to climate change and land degradation, including the adaptation of cropping and livestock systems, as well as ecosystem-based adaptation and SLM. SLM can address a number of the feedbacks between climate change and land degradation. For example, SLM may be able to help mediate the feedbacks between climate change and land degradation via changes in vegetation and soil carbon stocks. Rather than losing carbon due to land degradation, a number of SLM techniques are able to build soil organic matter and sequester significant amounts of carbon, thereby helping mitigate climate change. SLM practices also directly link to the feedback between climate change and land degradation that is mediated through losses of vegetation cover. Certain SLM technologies also have the potential to mitigate biodiversity-mediated feedbacks between climate change and land degradation.

There is a danger however, that adaptations based on scientific knowledge alone may not be suitable for the socio-cultural context in which they are needed, and this may significantly limit their uptake and effectiveness. By combining scientific understanding of adaptation options with local, contextual knowledge, it may be possible to develop adaptations based on generations' worth of experiential knowledge, which can help refine adaptations. It is therefore necessary to consider the benefits and drawbacks of both locally held and scientific knowledge for the development of adaptations to climate change and land degradation.

Finally, the chapter has considered how barriers to adaptation may be overcome. Barriers to adaptation may arise from a lack of available options to substitute one form of capital for another, low political capacity to enact strategies to support adaptation, high levels of institutional inertia and rigidity, lack of access to information about adaptation options, or financial constraints. Other barriers can be cognitive in nature, linked to a lack of perceived risk, a lack of perceived agency and a sense of powerlessness, low aspirations, or the social norms that influence behaviour within particular socio-cultural settings.

Once these barriers have been overcome, it is necessary to evaluate potential trade-offs between adaptations, so that complementary bundles of adaptations can be implemented together, avoiding maladaptation and reducing vulnerability to both climate change and land degradation. It is argued that ecosystem-based approaches and SLM have the potential to simultaneously enable adaptation to climate change and land degradation, whilst in many cases protecting or enhancing biodiversity. As such, SLM may be considered to offer potential for "triple-win" adaptation options.

## Notes

1 Adaptation is defined by IPCC (2014) as "reductions in risk and vulnerability through the actions of adjusting practices, processes and capital in response to the actuality or threat of climate change".
2 For example The European Climate Adaptation Platform (Climate-ADAPT) (http://climate-adapt.eea.europa.eu) or the UNCBD's Climate Adaptation Database (http://adaptation.cbd.int)
3 Defined as "the use of land resources including soil, water, animals and plants, for the production of goods to meet changing human needs, while simultaneously ensuring the long-term productive potential of these resources and ensuring their environmental functions" (Liniger and Critchley 2007: 10)
4 Defined as the "agronomic, vegetative, structural and/or management measures that prevent and control land degradation and enhance productivity in the field" (Liniger and Critchley 2007: 10)

## References

Adger, W.N., Arnell, N.W., Tompkins, E.L. 2005. Successful adaptation to climate change across scales. *Global Environmental Change* 15: 77–86.

Adger, W.N. 2006. Vulnerability. *Global Environmental Change* 16: 268–281.

Aguilera, E., Lassaletta, L., Gattinger, A., Gimeno, B.S. 2013. Managing soil carbon for climate change mitigation and adaptation in Mediterranean cropping systems: a meta-analysis. *Agriculture, Ecosystems & Environment* 168: 25–36.

Andersen, L., Kühn, B.F., Bertelsen, M., Bruus, M., Larsen, S.E., Strandberg, M. 2013. Alternatives to herbicides in an apple orchard, effects on yield, earthworms and plant diversity. *Agriculture, Ecosystems & Environment* 172: 1–5.

Aronson, J., Alexander, S. 2013. Ecosystem restoration is now a global priority: time to roll up our sleeves. *Restoration Ecology* 21: 293–296.

Bailey, I., Wilson, G.A. 2009. Theorising transitional pathways in response to climate change: technocentrism, ecocentrism, and the carbon economy. *Environment and Planning A*, 41: 2324–2341.

Barnett, J., O'Neil, S. 2010. Maladaptation. *Global Environmental Change* 20: 211–213.

Béné, C., Wood, R. G., Newsham, A. Davies, M. 2012. Resilience: new utopia or new tyranny? Reflection about the potentials and limits of the concept of resilience in relation to vulnerability reduction programmes. *IDS Working Papers* 2012: 1–61.

Berkes, F., Colding, J., Folke, C. 2000. Rediscovery of traditional ecological knowledge as adaptive management. *Ecological Applications* 10(5), 1251–1262.

Berkes, F., Folke, C. 2002. Back to the Future: Ecosystem dynamics and local knowledge. In: *Panarchy: understanding transformations in human and natural systems*, Gunderson, L.H., Holling, C.S. (eds). Island Press: Washington.

Biazin, B., Sterk G., Temesgen M., Abdulkedir, A., Stroosnijder, L. 2012. Rainwater harvesting and management in rainfed agricultural systems in sub-Saharan Africa – a review. *Physics and Chemistry of the Earth* 48: 139–151.

Bryant, A., Brainard, D.C., Haramoto, E.R., Szendrei, Z. 2013. Cover crop mulch and weed management influence arthropod communities in strip-tilled cabbage. *Environmental Entomology* 42: 293–306.

Burton, I., Diringer, E., Smith, J. 2006. *Adaptation to Climate Change: International policy options*. Pew Center on Global Climate Change: Arlington, VA, USA.

Ceccarelli, T., Bajocco, S., Salvati, L., Perini, L. 2014. Investigating syndromes of agricultural land degradation through past trajectories and future scenarios. *Soil Science and Plant Nutrition* 60: 1–11.

Compston, H. 2010. The politics of climate policy: strategic options for national governments. *Political Quarterly* 81: 107–115.

Cowie, A.L., Penman, T.D., Gorissen, L., Winslow, M.D., Lehmann, J., Tyrrell, T.D., Twomlow, S., Wilkes, A., Lal, R., Jones, J.W., Paulsch, A., Kellner, K., Akhtar-Schuster, M. 2011. Towards sustainable land management in the drylands: scientific connections in monitoring and assessing dryland degradation, climate change and biodiversity. *Land Degradation & Development* 22: 248–260.

De Bruin, K., Dellink, R.B., Ruijs, A., Boldwidt, L., van Buuren, A., Graveland, J., de Groot, R.S., Kuikman, P.J., Reinhard, S., Roetter, R.P., Tassone, V.C., Verhagen, A., van Ierland, E.C. 2009. Adapting to climate change in the Netherlands: an inventory of climate adaptation options and ranking of alternatives. *Climatic Change* 95: 23–45.

De Bruin, K.C., Dellink, R. 2011. How harmful are restrictions on adapting to climate change? *Global Environmental Change* 21: 34–45.

De Vente, J., Reed, M.S., Stringer, L.C., Valente, S., Newig, J. in press. How does the context and design of participatory decision-making processes affect their outcomes? Evidence from sustainable land management in global drylands. *Ecology & Society*.

Deressa, T., Hassan, R., Ringler, C., Alemu, T., Yesuf, M. 2009. Determinants of farmers' choice of adaptation methods to climate change in the nile basin of Ethiopia. *Global Environmental Change* 19: 248–255.

Dixon, J.L., Stringer, L.C., Challinor, A.J. 2014. Farming system evolution and adaptive capacity: insights for adaptation support. *Resources* 3: 182–214.

Dyer, J.C., Leventon, J., Stringer, L.C., Dougill, A.J., Syampungani, S., Nshimbi, M., Chama, F., Kafwifwi, A. 2013. Partnership models for climate compatible development: experiences from Zambia. *Resources* 2: 1–38.

Eakin, H., Lemos, M.C. 2006. Adaptation and the state: Latin America and the challenge of capacity-building under globalization. *Global Environmental Change* 16: 7–18.

Fazey, I., Fazey, J.A., Salisbury, J.G., Lindenmayer, D.B., Dovers, S. 2006. The nature and role of experiential knowledge for environmental conservation. *Environmental Conservation* 33: 1–10.

Fleskens, L., Nainggolan, D., Stringer, L.C. 2014. An exploration of scenarios to support sustainable land management using integrated environmental socio-economic models. *Environmental Management* 54: 1005–1021.

Forsyth, T. 1996. Science, myth and knowledge: testing Himalayan environmental degradation in Thailand. *Geoforum* 27(3): 375–392.

Freier, K.P., Bruggemann, R., Scheffran, J., Finckh, M. Schneider, U.A. 2012. Assessing the predictability of future livelihood strategies of pastoralists in semi-arid Morocco under climate change. *Technological Forecasting and Social Change* 79: 371–382.

Friedkin, N.E. 1998. *A Structural Theory of Social Influence*. Cambridge University Press: New York.

Fujisawa, M.K. Koyabashi, K. 2010. Apple (*Malus pumila var. domestica*) phenology is advancing due to rising air temperature in northern Japan. *Global Change Biology* 16: 2651–2660.

Füssel, H.M. 2009. An updated assessment of the risks from climate change based on research published since the IPCC Fourth Assessment Report. *Climatic Change* 97: 469–482.

Garbach, K., Lubell, M., DeClerck, A.J. 2012. Payment for ecosystem services: the roles of positive incentives and information sharing in stimulating adoption of silvopastoral conservation practices. *Agriculture, Ecosystems & Environment* 156: 27–36.

Goodman-Elgar, 2008. Evaluating soil resilience in long-term cultivation: a study of pre-Columbian Terraces from the Paca Valley, Peru. *Journal of Archaeological Science* 35: 3072–3086.

Grainger, A. 2014. Is land degradation neutrality feasible in dry areas? *Journal of Arid Environments* 112: 14–24.

Homans, G.C. 1950. *The Human Group*. Routledge and Kegan Paul: London.

Howden, S., Crimp, S., Nelson, R. 2010. Australian agriculture in a climate of change. In: *CSIRO Publishing Proceedings of Managing Climate Change: Papers from the GREENHOUSE 2009 Conference*, Jubb, I.P. Holper, P., Cai, W. (eds). CSIRO: Melbourne, pp. 101–111.

Huntjens, P., Pahl-Wostl, C., Grin, J. 2010. Climate change adaptation in European river basins. *Regional Environmental Change* 10: 263–284.

IPCC (International Panel of Climate Change). 2001. *Climate Change 2001: Impacts, Adaptation, and Vulnerability*. Cambridge University Press: Cambridge.

IPCC. 2007. Climate Change 2007: The Physical Science Basis. In: *Contribution of Working Group I to the fourth assessment report of the intergovernmental panel on climate change*, Solomon, S., Qin, H.D., Manning, M., Chen, Z., Marquis, M., Averyt, K.B., Tignor, M., Miller, H.L. (eds). Cambridge University Press: Cambridge, United Kingdom and New York, NY, USA.

IPCC. 2014. Climate Change 2014. Impacts, Adaptation, and Vulnerability. In: *Part A: global and sectoral aspects. Contribution of Working Group II to the fifth assessment report of the intergovernmental panel on climate change*, Field, C.B., Barros, V.R., Dokken, D.J., Mach, K.J., Mastrandrea, M.D., Bilir, T.E., Chatterjee, M., Ebi, K.L., Estrada, Genova, Y.O., Girma, R.C., Kissel, B., Levy, E.S., MacCracken, A.N., Mastrandrea,

P.R., White, L.L. (eds). Cambridge University Press: Cambridge, UK and New York, NY, USA.

Jones, H.P., Hole, D.G., Zavaleta, E.S. 2012. Harnessing nature to help people adapt to climate change. *Nature Climate Change* 2: 504–509.

Kabubo-Mariara, J. 2009. Global warming and livestock husbandry in Kenya: impacts and adaptations. *Ecological Economics* 68: 1915–1924.

Kalaba, F.K., Quinn, C.H., Dougill, A.J. 2014. Policy coherence and interplay between Zambia's forest, energy, agricultural and climate change policies and multilateral environmental agreements. *International Environmental Agreements: Politics, Law and Economics* 14: 181–198.

Kalanda-Joshua, M., Ngongondo, C., Chipeta, L., Mpembeka, F. 2011. Integrating indigenous knowledge with conventional science: enhancing localised climate and weather forecasts in Nessa, Mulanje, Malawi. *Physics and Chemistry of the Earth, Parts A/B/C* 36: 996–1003.

Klein, R.J.T., Midgley, G.F., Preston, B.L., Alam, M., Berkhout, F., Dow, K., Shaw, M.R. 2014. Adaptation opportunities, constraints, and limits. In: *Climate Change 2014: Impacts, adaptation, and vulnerability. Part A: Global and sectoral aspects. Contribution of Working Group II to the fifth assessment report of the intergovernmental panel on climate change*, Field, C.B., Barros, V.R., Dokken, D.J., Mach, K.J., Mastrandrea, M.D., Bilir, T.E., Chatterjee, M., Ebi, K.L., Estrada, Y.O., Genova, R.C.,. Girma, B, Kissel, E.S., Levy, A.N., MacCracken, S., Mastrandrea, P.R., White, L.L. (eds). Cambridge University Press: Cambridge, United Kingdom and New York, NY, USA, pp. 899–943.

Lal, R. 2004. Carbon sequestration in dryland ecosystems. *Environmental Management* 33: 528–544.

Lal, R. 2007. Carbon management in agricultural soils. *Mitigation and Adaptation Strategies for Global Change* 12: 303–322.

Lapeyre, R., Pirard, R. 2013. Payments for environmental services and market-based instruments: next of kin or false friends? *IDDRI Working Paper 14/13, Institute for Sustainable Development and International Relations*. IDDRI: Paris, 16pp.

Licznar-Małańczuk, M. 2014. The diversity of weed species occurring in living mulch in an apple orchard. *Acta Agrobotanica* 67: 47–54.

Liniger, H.P., Critchley, W. (eds). 2007. *Where the Land is Greener – Case Studies and Analysis of Soil and Water Conservation Initiatives Worldwide*. WOCAT. CTA: Wageningen.

McDowell, J.J., Hess, J. 2012. Accessing adaptation: multiple stressors on livelihoods in the Bolilvian Highlands under a changing climate. *Global Environmental Change* 22: 342–352.

McGuire, S.J., Sperling, L. 2008. Leveraging farmers' strategies for coping with stress: seed aid in Ethiopia. *Global Environmental Change-Human and Policy Dimensions* 18: 679–688.

McKeon, G.M., Stone, G.S., Syktus, J.I., Carter, J.O., Floof, N.R., Ahrens, D.G., Bruget, D.N., Chilcott, C.R., Cobon, D.H., Cowley, R.A., Crimp, S.J., Fraser, G.W., Howden, S.M., Johnston, P.W., Ryan, J.G., Stokes, C.J., Day, K.A. 2009. Climate change impacts on northern Australian rangeland livestock carrying capacity: a review of issues. *The Rangeland Journal* 31: 1–29.

MA (Millennium Ecosystem Assessment). 2005. *Ecosystems and Human Well-being: Policy responses*. Island Press: Washington, D.C., USA.

Mazzucato, V.D. Niemeijer, 2000. The cultural economy of soil and water conservation: market principles and social networks in eastern Burkina Faso. *Development and Change* 31: 831–855.

Mekdaschi Studer, R., Liniger, H. 2013. *Water Harvesting: Guidelines to good practice*. Centre for Development and Environment (CDE), Bern; Rainwater Harvesting Implementation

Network (RAIN), Amsterdam; MetaMeta, Wageningen; The International Fund for Agricultural Development (IFAD), Rome.

Mertz, O., Mbow, C., Reenberg, A., Diouf, A. 2009. Farmers' perceptions of climate change and agricultural adaptation strategies in rural Sahel. *Environmental Management* 43: 804–816.

Oba, G., Kotile, D.G. 2001. Assessments of landscape level degradation in southern Ethiopia: pastoralists versus ecologists. *Land Degradation & Development* 12(5): 461–475.

Ogden, A.E., Innes, J.L. 2009. Adapting to climate change in the southwest Yukon: locally identified research and monitoring needs to support decision making on sustainable forest management. *Arctic* 62: 159–174.

Olesen, J., Tranka, M., Kersebaum, K., Skjelvag, A., Seguin, B., Peltonen-Sainio, P.F., Rossi, F., Kozyra, J., Micale, F. 2011. Impacts and adaptation of European crop production systems to climate change. *European Journal of Agronomy* 34: 96–112.

Olsson, P., Folke, C. 2001. Local ecological knowledge and institutional dynamics for ecosystem management: a study of Lake Racken watershed, Sweden. *Ecosystems* 4: 85–104.

Pelling, M., High, C., Dearing, J., Smith, D. 2008. Shadow spaces for social learning: a relational understanding of adaptive capacity to climate change within organisations. *Environment and Planning A* 40: 867–884.

Pinkse, J., Kolk, A. 2012. Addressing the climate change – sustainable development nexus. *Business and Society* 51: 176–210.

Pirard, R., Billé, R., Sembrés, T. 2010. Questioning the theory of payments for ecosystem services (PES) in light of emerging experience and plausible developments. *Institut du Développement Durable et des Relations Internationales.* Paris.

Pittock, J. 2011. National climate change policies and sustainable water management: conflicts and synergies. *Ecology and Society* 16: 25.

Prell, C., Hubacek, K., Reed, M.S. 2009. Social network analysis and stakeholder analysis for natural resource management. *Society & Natural Resources* 22: 501–518.

Quinn, C.H., Ziervogel, G., Taylor, A., Takama, T., Thomalla, F. 2011. Coping with multiple stresses in rural South Africa. *Ecology and Society* 16: 2.

Raymond, C.M., Fazey, I., Reed, M.S., Stringer, L.C., Robinson, G.M., Evely, A.C. 2010. Integrating local and scientific knowledge for environmental management: from products to processes. *Journal of Environmental Management* 91: 1766–1777.

Reed, M.S. 2008. Stakeholder participation for environmental management: a literature review. *Biological Conservation* 141: 2417–2431.

Reed, M.S., Buenemann, M., Atlhopheng, J., Akhtar-Schuster, M., Bachmann, F., Bastin, G., Bigas, H., Chanda, R., Dougill, A.J., Essahli, W., Evely, A.C., Fleskens, L., Geeson, N., Glass, J.H., Hessel, R., Holden, J., Ioris, A., Kruger, B., Liniger, H.P., Mphinyane, W., Nainggolan, D., Perkins, J., Raymond, C.M., Ritsema, C.J., Schwilch, G., Sebego, R., Seely, M., Stringer, L.C., Thomas, R., Twomlow, S., Verzandvoort, S. 2011. Cross-scale monitoring and assessment of land degradation and sustainable land management: a methodological framework for knowledge management. *Land Degradation & Development* 22: 261–271.

Reed, M.S., Podesta, G., Fazey, I., Beharry, N.C., Coen, R., Geeson, N., Hessel, R., Hubacek, K., Letson, D., Nainggolan, D., Prell, C., Psarra, D., Rickenbach, M.G., Schwilch, G., Stringer, L.C., Thomas, A.D. 2013a. Combining analytical frameworks to assess livelihood vulnerability to climate change and analyse adaptation options. *Ecological Economics* 94: 66–77.

Reed, M.S., Hubacek, K., Bonn, A., Burt, T.P., Holden, J., Stringer, L.C., Beharry-Borg, N., Buckmaster, S., Chapman, D., Chapman, P., Clay, G.D., Cornell, S., Dougill, A.J., Evely, A., Fraser, E.D.G., Jin, N., Irvine, B., Kirkby, M., Kunin, W., Prell, C., Quinn, C.H.,

Slee, W., Stagl, S., Termansen, M., Thorp, S., Worrall, F. 2013b. Anticipating and managing future trade-offs and complementarities between ecosystem services. *Ecology & Society* 18(1): 5.

Reed, M.S., Stringer, L.C., Dougill, A.J., Perkins, J.S., Atlhopheng, J.R., Mulale, K., Favretto, N. 2015. Reorienting land degradation towards sustainable land management: linking sustainable livelihoods with ecosystem services in rangeland systems. *Journal of Environmental Management* 151: 472–485.

Reid, P., Vogel, C., 2006. Living and responding to multiple stressors in South Africa – glimpses from KwaZulu-Natal. *Global Environmental Change* 16: 195–206.

Rockström, J., Hatibu, N., Oweis, T.Y., Wani, S., Barron, J., Bruggeman, A., Farahani, J., Karlberg, L. and Z. Qiang. 2007. Managing water in rainfed agriculture. In: *Water for Food, Water for Life: Comprehensive assessment of water management in agriculture*, Molden, D. (ed.). Earthscan and International Water Management Institute (IWMI): London and Colombo.

Rogers, E.M. 1995. *Diffusion of Innovations*, 4th edition. The Free Press: New York.

Ruef, M., Aldrich, H.E., Carter, N.M. 2003. The structure of founding teams: homophily, strong ties, and isolation among U.S. entrepreneurs. *American Sociological Review* 68: 195–222.

Schneider, S.H., Easterling, W.E., Mearns, L.O. 2000. Adaptation: sensitivity to natural variability, agent assumptions and dynamic climate changes. *Climatic Change* 45: 203–221.

Scopel, E., Triomphe, B., Affholder, F., Da Silva, F.A.M., Corbeels, M., Xavier, J.H.V., De Tourdonnet, S. 2013. Conservation agriculture cropping systems in temperate and tropical conditions, performances and impacts. A review. *Agronomy for sustainable development* 33: 113–130.

Seely M., Montgomery S. 2011. Proud of our deserts: combating desertification – an NGO perspective on a national programme to combat desertification. Available online at: www.drfn.org.na/projects/old-projects/napcod

Silvestri, S., Bryan, E., Ringler, C., Herrero, M., Okoba, B. 2012. Climate change perception and adaptation of agro-pastoral communities in Kenya. *Regional Environmental Change* 12: 791–802.

Simelton E., Quinn C.H., Batisani N., Dougill A.J., Dyer J., Fraser E., Mkwambisi D., Sallu S.M., Stringer L. 2013. Is rainfall really changing? Farmers' perceptions, meteorological data, and policy implications. *Climate and Development* 5: 123–138.

Simon, H.A. 1955. A behavioral model of rational choice. *Quarterly Journal of Economics* 69: 99–118.

Simon, H.A. 1956. Rational choice and the structure of the environment. *Psychological Review* 63: 129–138.

Skvoretz, J., Fararo, T.J., Agneesens, F. 2004. Advances in biased net theory: definitions, derivations, and estimations. *Social Networks* 26: 113–139.

Smit, B., Wandel, J. 2006. Adaptation, adaptive capacity and vulnerability. *Global Environmental Change* 16: 282–292.

Smith, J.B., Lenart, S.S. 1996. Climate change adaptation policy options. *Climate Research* 6: 193–201.

Smith, J.B., Vogel, J.M., Cromwell, J.E., III. 2009. An architecture for government action on adaptation to climate change. *Climatic Change* 95: 53–61.

Speranza, C.I., Kiteme, B., Ambenje, P., Wiesmann, U., Makali, S. 2010. Indigenous knowledge related to climate change variability and change: insights from droughts in semi-arid areas of former Makueni Distict, Kenya. *Climate Change* 100: 295–315.

Stave, J., Oba, G., Nordal, I., Stenseth, N.C. 2007. Traditional ecological knowledge of a riverine forest in Turkana, Kenya: implications for research and management. *Biodiversity and Conservation* 16: 1471–1489.

Stringer, L.C., Reed, M.S. 2007. Land degradation assessment in southern Africa: integrating local and scientific knowledge bases. *Land Degradation and Development* 18: 99–116.

Stringer L.C., Dyer, J., Reed, M., Dougill, A., Twyman, C., Mkwambisi, D. 2009. Adaptations to climate change, drought and desertification: insights to enhance policy in southern Africa. *Environmental Science and Policy* 12: 748–765.

Stringer, L.C., Fleskens, L., Reed, M.S., de Vente, J., Zengin, M. 2014. Participatory evaluation of monitoring and modeling of sustainable land management technologies in areas prone to land degradation. *Environmental Management* 54: 1022–1042.

Stringer, L.C., Quinn, C.H., Berman, R., Dixon, J.L. In press. Livelihood adaptation and climate variability in Africa. In: *Handbook of International Development*, Hammett, D., Grugel, J. (eds). Palgrave, UK.

Suckall, N., Tompkins, E., Stringer, L.C. 2014. Identifying trade-offs between adaptation, mitigation and development in community responses to climate and socio-economic stresses: evidence from Zanzibar, Tanzania. *Applied Geography* 46: 111–121.

Thomalla, F., Downing, T., Spanger-Siegfried, E., Han G.Y., Rockstrom, J. 2006. Reducing hazard vulnerability: toward a common approach between disaster risk reduction and climate adaptation. *Disasters* 30: 39–48.

Thomas, D.S.G., Twyman, C. 2004. Good or bad rangeland? Hybrid knowledge, science, and local understandings of vegetation dynamics in the Kalahari. *Land Degradation and Development* 15: 215–231.

Thomas, D.S.G., Knight, M., Wiggs, G.F.S. 2005. Remobilization of southern African desert dune systems by twenty-first century global warming. *Nature* 435: 1218–1221.

Thomas, R.J. 2008. 10th Anniversary review: Addressing land degradation and climate change in dryland agroecosystems through sustainable land management. *Journal of Environmental Monitoring* 10: 595–603.

Tol, R., Fankhauser, S., Smith, J.B. 1998. The scope for adaptation to climate change: what can we learn from the impact literature? *Global Environmental Change* 8: 109–123.

Van Aalst, M.K., Cannon, T., Burton, I. 2008. Community level adaptation to climate change: the potential role of participatory community risk assessment. *Global Environmental Change* 18: 165–179.

Wilson, G.A. 2014. Community resilience: path dependency, lock-in effects and transitional ruptures. *Environment and Planning A* 57: 1–26.

World Bank 2010. *World Development Report: Development in a changing climate – concept note.* World Bank, Washington, D.C., USA, 45pp.

Yohe, G., Tol, R.S.J. 2001. Indicators for social and economic coping capacity – moving toward a working definition of adaptive capacity. *Global Environmental Change* 12: 25–40.

Ziervogel, G., Bharwani, S., Downing, T.E. 2006. Adapting to climate variability: pumpkins, people and policy. *Natural Resources Forum* 30: 294–305.

# 7

# MONITORING AND EVALUATING CURRENT AND FUTURE EFFECTS OF CLIMATE CHANGE AND LAND DEGRADATION

This chapter considers how best to monitor and evaluate interventions to enhance the capacity of ecosystems and populations to adapt to climate change and land degradation. Decision-makers, from landowners and managers, to regional, national and international stakeholders, need to be able to effectively evaluate response options and monitor their success or failure, so that responses can be refined in future. This is important because the complex linkages within and between social-ecological systems mean that the effects of different responses may have unexpected consequences for linked ecosystems and populations, and the ecosystem services upon which they depend. When evaluating the suitability and effectiveness of adaptations to both climate change and land degradation, it is important to consider likely effects on ecosystem processes and ecosystem services, and how these then impact upon livelihoods.

It is also essential to assess the socio-cultural context in which these adaptations might be used, for example, considering the skills and resources required to implement them, their cultural acceptability and their compatibility with existing institutional arrangements, such as land tenure systems. Understanding the likely consequences of different response options is highly complex, and can only be done in collaboration with those who may use the options. For this reason, cooperation between researchers and policy makers, together with practitioners and local communities, is vital if we are to fully consider the likely implications of different response options and to be able to appropriately use monitoring data to refine future responses.

This chapter starts by reviewing approaches for monitoring and evaluating the interactive effects of climate change and land degradation, considering both current (Section 7.1) and likely future effects (Section 7.2).

## 7.1 Monitoring and evaluating current effects of land degradation and climate change

Following the conceptual framework described in Chapter 3, where ecosystems and populations are exposed and sensitive to climate change and land degradation, there are a number of potential interactions that may occur between these two processes that can impact upon livelihoods and human well-being. Understanding these sensitivities is essential to identify appropriate adaptations that may reduce the vulnerability of these ecosystems and populations to the interactive effects of climate change and land degradation. Monitoring and evaluation is also essential to determine the effectiveness of adaptations so that they can be refined where necessary to enhance resilience. As described in the methodological framework in Chapter 3, there are three broad approaches to monitoring: (i) direct measurements; (ii) proxy measurements or indicators; and (iii) model-based approaches. Each of these approaches has a number of benefits and drawbacks.

Direct measurements are the most accurate approach, but can be extremely costly and time-consuming. The accuracy of direct measurements makes it possible to reliably compare ecosystem processes and the provision of ecosystem services between locations and over time, providing a detailed account of changes as they occur, with the ability to interpret the likely causes of those changes via understanding of the underlying processes driving change (see Box 7.1 for details of direct measurement options for soil carbon). Although sampling regimes are typically used to represent systems and reduce the number of measurements required, the level of investment that is needed is typically beyond the reach of landowners and managers, and not feasible at scales relevant to regional, national or international stakeholders. For example, the heterogeneity of soil characteristics typically requires a high sampling frequency, and laboratory testing is necessary to accurately measure many of these characteristics. Similarly, directly measuring the impacts of climate change and land degradation on livelihoods typically requires household surveys, which may be feasible at a village scale, but which become time-consuming and costly at broader spatial scales. For these reasons, proxy measurements or indicators are often used to represent changes in ecosystem processes and services, and assess their likely impacts on livelihoods. By definition, indicators will only ever be an approximation or indication of change, and may sometimes provide misleading guidance for decision-makers. It is therefore typically necessary to monitor and triangulate data between a range of different indicators to reliably assess the effects of climate change and land degradation (Reed et al., 2006, 2011).

As remotely sensed products become increasingly available (e.g. GlobCover, Afri-Cover), geospatial approaches are being used to monitor and assess various aspects of land degradation and climate change. Most work has focused on quantifying levels and changes in land cover, biological productivity, and water quality and quantity. Land cover maps are commonly used in assessments of, for example, soil erosion (Beskow et al., 2009), species habitat suitability (Cecchi et al., 2008), or water pollution risk (Backhaus et al., 2002). Global Net Primary Production data have been routinely available since the early 1980s, and have been used widely to indicate land degradation or improvement (e.g. Running et al. (2004) as an example

# BOX 7.1: OPTIONS FOR DIRECT MEASUREMENTS OF SOIL CARBON

Direct measurements of soil carbon (whether organic or inorganic) can provide important information about land degradation processes and the climate adaptation and mitigation potential of land. There are four main soil carbon measurement methods: (i) wet oxidation (such as the Walkley-Black method, with the Anne method being the French variant); (ii) combustion methods to determine the $CO_2$ produced (e.g. IR, titration, conductimetry); (iii) near-infrared spectroscopy (NIRS); and (iv) laser-induced breakdown spectroscopy (LIBS).

Wet oxidation methods directly measure organic carbon following organic matter oxidation via excess potassium bichromate in sulphuric acid at 135°C. The quantity of chrome III+ formed, proportional to the soil organic carbon content, is determined by colorimetry. However, oxidation may be incomplete, which means that only part of the organic carbon is extracted, which is often the case in tropical or carbonate-rich soils (Bernoux and Chevallier, 2014).

Combustion methods are more commonly used to determine the total soil carbon (organic and inorganic). The method described in the NF ISO 10694 standard involves micro-weighing (around 25 mg), flash combustion, chromatographic separation of molecular nitrogen and carbon dioxide, and thermal conductivity detection. It is essential to know the inorganic carbon content from the outset in order to determine the organic carbon content, otherwise the sample has to be decarbonated prior to analysis (Bernoux and Chevallier, 2014).

Over the past two decades, visible – near infrared (vis – NIR) diffuse reflectance spectroscopy has been increasingly used to determine soil organic and inorganic carbon. To generate a soil spectrum, radiation containing all relevant frequencies in a particular range is directed at the sample. Depending on the constituents present in the soil, the radiation will cause individual molecular bonds to vibrate, either by bending or stretching, and they will absorb light, to various degrees, with a specific energy quantum corresponding to the difference between two energy levels. As the energy quantum is directly related to frequency (and inversely related to wavelength), the resulting absorption spectrum produces a characteristic shape that can be used for analytical purposes. The use of vis – NIR involves only drying and crushing samples, without the need for (hazardous) chemicals, and measurement takes only a few seconds (Stenberg et al., 2010).

Laser-Induced Breakdown Spectroscopy (LIBS) uses a pulsed laser to heat the soil sample to a temperature of 1350°C in the presence of oxygen. Combustion converts soil organic matter into $CO_2$, which can be measured using infrared spectroscopy (Martin et al., 2003).

Adapted from Bernoux and Chevallier (2014), Stenberg et al. (2010) and Martin et al. (2003).

of the use of coarse scale resolution data and Ollinger and Smith (2005) for finer spatial resolutions for local to national scale assessments).

Multi-temporal GIS analysis of land cover and land capability dynamics together with the use of landscape metrics (general or specifics) may help us to understand patterns and structures of land degradation, and identify the loss of cultural land-scapes. Also, the development of such "permanent" evaluation techniques can help in determining environmental thresholds at which further land cover change should be prevented (e.g. Pascual Aguilar *et al.*, 2011). Remote sensing, and geospatial technologies in general, are powerful tools to provide for initial assessments (or development of baselines) at multiple scales related to land degradation pro-cesses, including vulnerability to climate change assessments, mapping of hotspots (areas for intervention) and bright spots (i.e. examples of good policy/strategies) (Metternicht, 2014). Such approaches can assist in capturing the temporal and spa-tial dynamics of vulnerability and adaptive capacities of ecosystems. For example, the UNEP's REGATTA (Regional Gateway for Technology Transfer and Action on Climate Change in Latin America and the Caribbean) project implemented in the Gran Chaco Americano used GIS to generate maps of ecoregions, land cover change, and regulating, supporting and provisioning ecosystem services. This approach was combined with stakeholder consultations to identify the threats, status and trends of the ecosystem services in the context of current and future climate conditions (Metternicht *et al.*, 2014). Similarly, the "Mapping Hotspots of Climate Change and Food Insecurity in the Global Tropics" project aimed to iden-tify areas that are food insecure and vulnerable to the impacts of climate change. The study used maps of variables that indicate different aspects of food security (availability, access and utilization) and considered the thresholds of climate change exposure important for agricultural systems. Vulnerability was assessed as a function of exposure, sensitivity and coping capacity. Tropical areas of interest were classi-fied by high or low exposure, high or low sensitivity and high or low coping capac-ity. Such spatially explicit representations can help in identifying priority areas for in-depth monitoring and assessment (see https://cgspace.cgiar.org/handle/10568/3826).

Buenemann *et al.* (2011) argue that approaches like these should be used as part of a suite of indicators linking human and environmental data, qualitative and quantitative data, as well as field and remotely sensed data in a spatially explicit framework. This "integrative geospatial approach" has the potential to represent different stakeholder perspectives and promote knowledge exchange between diverse actors.

Nevertheless, it has been claimed that indicators tend to provide few benefits to landowners and managers, who as a consequence rarely apply them (Carruthers and Tinning, 2003; Innes and Booher, 1999). Partly, this is because indicators are usu-ally developed by experts, and applied without engaging effectively with local communities and other decision-makers (Riley, 2001). However, the UNCCD stresses the need for local communities to participate in all stages of project plan-ning and implementation, including the selection, collection and monitoring of indicators (WCED, 1987; UNCCD, 1994; Corbiere-Nicollier *et al.*, 2003). To do this, the methods used to collect, apply and interpret indicators must be easily used by non-specialists. To achieve widespread uptake, indicators must also be clearly linked to the needs, priorities and goals of the decision-makers who need to use them. In the hands of local communities, and regional and national decision-makers affected by climate change and land degradation, indicators have the potential to enhance the overall understanding of environmental and social challenges and empower decision-makers to respond appropriately without having to rely on external experts (Reed *et al.*, 2006). Indicators have the potential to provide spatial comparisons if a core set of indicators can be used to compare progress in different locations. This can be done at local and international scales, though at these broader spatial scales, indicators tend to be based more on scientific rather than local knowledge (Reed *et al.*, 2011).

Such a comparative approach is useful if indicators are being used to assess progress towards combating DLDD. At COP11, the UNCCD adopted a set of six "progress indicators" (two indicators for each of the convention's strategic objectives), which will be used in reporting to the convention in 2016 (via the performance review and assessment of implementation system, PRAIS) and linked to NAPs (Decision 22/COP.11). The approach uses readily available data and attempts to link global level reporting with monitoring data from local and national scales, and combines qualitative and quantitative elements. It also emphasizes stakeholder participation, and integrates effects on human well-being with effects on ecosystem services using the DPSheIR (Driving Force, Pressure, State, human and environmental Impact and Response) framework.

Regional scale studies typically use modelling approaches. However, it is important to combine modelling with measured data from regional scales too. While experimental design is often difficult at regional scales, there are various innovative methods available to make regional scale assessments, including, for instance tracer studies and remote sensing techniques (e.g. Boix-Fayos *et al.*, 2014; Vanacker *et al.*, 2014). Model-based approaches have the potential to assess relationships between multiple variables, many of which may be indicators, checked or "validated" against direct measurements. As such, models are particularly well-suited to assessing the likely interactions between climate change and land degradation. Section 7.2 considers how model-based approaches compare to alternative approaches for assessing likely future climate change and land degradation.

In reality, a combination of direct measurements, indicators and models is likely to be required to understand the complex interactions between climate change and land degradation and monitor their effects. Reed *et al.* (2011) and Hessel *et al.* (2014) suggest a hybrid framework which was applied internationally through the EU-funded DESIRE project, building on approaches developed by the FAO's Land Degradation Assessment in Drylands (LADA), the WOCAT programme and the Dryland Development Paradigm (DDP). Although focused on monitoring land degradation and SLM, the framework is equally applicable to monitoring the effects of climate change, and the interactions between climate change and land degradation.

Following this framework, indicators would be developed and used to monitor climate change, land degradation and adaptation responses, enabling local communities and regional decision-makers to collect, analyse and act on monitoring results. Models would then be used to upscale assessments to national and international scales, to inform decision-makers at this scale about likely challenges and to inform the development of policy response options. Direct measurements would be used in the development and testing of indicators, and to calibrate models to new context and validate their outputs.

## 7.2 Assessing likely future effects of climate change and land degradation

The complex and uncertain interactions that are likely to take place between land degradation and climate change make it difficult to predict what the future may hold for the parts of the world that will be affected. However, to develop appropriate responses in policy and in practice, it is necessary to understand the type, direction and magnitude of the challenges that these processes will create. These responses might be as much about harnessing benefits from the challenges posed by land degradation and climate change; as Louis Pasteur said, "chance favours only the prepared minds". Broadly, there are three ways we can anticipate the future and set out policies and strategies that can move us closer to the future we want: prediction, visioning and scenarios (Reed *et al.*, 2013). This section considers the benefits and drawbacks of each of these approaches in the context of anticipating likely interactions between climate change and land degradation.

### 7.2.1 *Predictive approaches*

Through history, people have effectively used cues from their environment to predict environmental change over relatively short timescales. This includes, for example, predicting the weather from perceptible declines in atmospheric pressure or storm clouds on the horizon, or predicting the onset of rains after drought before any storm clouds are visible by looking for buds on certain species of trees. However, it is only relatively recently that mathematical models have enabled us to predict anything over longer time horizons and to assign confidence levels to our projections. Now with increasing computer power, these models are capable of depicting ever more complex processes.

For example, IPCC (2014) created 900 mitigation scenarios based on large-scale, integrated models, including a range of technological, socioeconomic and institutional trajectories. Detailed outputs from these models and their links to land degradation were reviewed earlier in this book. These models link human systems (including energy, land use and economy) with the physical processes of climate change, to identify cost-effective mitigation outcomes. However, IPCC (2014:10) warns that "they are simplified, stylized representations of highly complex, real-world processes, and the scenarios they produce ... can differ considerably from the reality that unfolds".

All models are essentially simplified representations of reality. They seek to represent real-world processes as logically and realistically as possible, but ultimately are abstractions of reality. Given the huge range of possible variables, interactions and feedback loops in any socio-ecological system, models concerned with climate change and land degradation processes have to focus on the most important components of the system that is being modelled. These would typically be the aspects that explain most of the change or variability that the system exhibits. In this sense, building a model is a bit like painting a portrait. A skilled portrait painter can represent their subject deftly with a few brush strokes, and although somewhat abstracted from reality when compared to a photograph, the subject is instantly recognizable to the viewer. The painter could add many more brush strokes to more accurately represent the subject's face, but it is unlikely to do much to increase their recognizability. The more abstract the portrait, the wider the variety of ways it can be interpreted by those who view it.

Model builders face the same challenge: to build up their "portrait" of the system with no more components and interactions than are necessary to represent the way it functions. However, if the model becomes too abstract or simplified, it may not adequately represent the complexity of the real system. When this happens, it can sometimes mean that the results are misleading. As different modellers focus on different aspects of the system, and represent processes in different ways, discrepancies appear between competing models. These discrepancies tend to fuel uncertainty, and can cloud decision making.

Often, future projections are sought using increasingly powerful computer models that couple or integrate elements of socio-economic and environmental systems (Prell *et al.*, 2007). Partly, the goal is to more precisely represent real-world systems, where people are part of the environment in which they live – like a portrait painter zooming out from their subject to depict them as part of a landscape or event, for example riding horseback through a landscape or as part of a battle scene. But these integrated models also have the potential to help us understand how humans are likely to behave in response to environmental change, and in turn understand how their actions will alter the course of those environmental changes.

There is evidence that people are already adapting to the twin challenges of climate change and land degradation and have been doing so for millennia (e.g. Stringer *et al.*, 2009), and these adaptations will themselves help to mediate the consequences of climate change and land degradation. Only by understanding how people are likely to react to these challenges, can we realistically anticipate the nature of the challenges we will face. However, by casting the net this wide, the number of potential variables that must be considered starts to mushroom, and difficult decisions must be made about where to focus modelling activity. To use the portrait example again, rather than just trying to depict a single person, it is like trying to accurately render an entire city of people's faces in a single portrait along with their interactions with each other and surroundings. Some level of abstraction and simplification is inevitable.

Ultimately, given the enormous complexity and dynamism of coupled socio-economic and environmental systems, computer models can offer a number of very *precise* insights into the ways that they may respond to climate change and land degradation, but at the same time, they may be precisely wrong. With their numerical outputs, statistical probabilities and maps, there is a risk that such models present an illusion of certainty to decision-makers. The danger of this is that those who attempt to use predictive models to prepare for the future focus on adapting only for the single future that the model predicted. In doing so, they may neglect to prepare for a range of equally plausible futures that may in fact come to pass. Therefore, if the model turns out to be wrong, they find themselves unprepared and less likely to be able to adapt rapidly and effectively for the future in which they find themselves.

## 7.2.2 Visioning approaches

An alternative to predicting the effects of climate change and land degradation is to envision idealized futures and then consider how these futures might be realized in the context of climate change and land degradation. "Visioning" and "back-casting" exercises are an increasingly popular way for decision-makers to prepare for the future (Wilson, 1992; Dreborg, 1996; Manning *et al.*, 2006). These involve a structured, group process of envisioning desirable future states, and then working back to the present day, to think about the actions or conditions that would be necessary to achieve the vision.

Although visioning processes can be a powerful and creative way of getting diverse groups of people to plan for the future together, some groups within society have more power than others to pursue their vision for the future, which inevitably raises issues of equity and distributional justice (Konow, 2003). There is also another problem: most of us are simply not that ambitious or imaginative. This is known as the "status quo bias". When asked what they would like the future to look like, most people look around themselves, and say "this" (Samuelson and Zeckhauser, 1988). There are a few people who have radical visions for the future, but since they are in a minority, their vision is likely to be unpopular with the majority. For example, many have argued that marginal agricultural land should be abandoned (or at least managed less intensively) as a way of reversing land degradation and increasing resilience to climate change (e.g. Scherr, 1999). However, proponents of re-wilding are deeply unpopular with land managers and many others who value the ecosystem services currently associated with agricultural landscapes. Some people warn that intensification of agriculture and climate change may limit that extent to which abandoned land can return to its original vegetation state (Cramer *et al.*, 2008), and others argue that land abandonment is in fact a major form of land degradation in its own right (e.g. Cerdà, 1997; Bajocco *et al.*, 2012). There are also fears relating to impacts on wildfires and the spread of invasive species (Benayas *et al.*, 2007; Cramer *et al.*, 2008).

### 7.2.3 Scenario-based approaches

Scenarios represent an alternative to prediction and visioning. As Kay (1989) suggests, "the best way to predict the future is to invent it". This may be as much about anticipating "nightmare" scenarios that no-one would want to happen as it is about envisioning ideal futures we want to pursue. By developing scenarios, we can explore different people's visions of the future and be prepared for whatever happens. Scenario development (or scenario "analysis" or "planning") is a therefore a "systematic method for thinking creatively about dynamic, complex and uncertain futures, and identifying strategies to prepare for a range of possible outcomes" (Reed *et al.*, 2013: 346).

The wider the range of different plausible scenarios we develop and prepare for, the more likely we are to prepare for something close to what actually happens and be able to adapt effectively. Even if the future turns out to be quite different, the process of thinking about how we might adapt to a range of different futures may still help us adapt more effectively than if we had not prepared. For example, by preparing for the effects of gradual land abandonment or re-wilding of marginal agricultural land, we may also be prepared for a range of other futures that have similar effects, for example the emergence of a new, untreatable animal or crop disease that leads to much more rapid land abandonment than we had previously anticipated.

Scenarios concerned with climate change and land degradation are often the domain of modellers, who try and understand how different scenarios might play out on the basis of their process-based understanding of the systems under consideration. Scenarios may however combine both quantitative and qualitative information, including the hopes and fears of the people who live and work in the system that is being studied. There is a normative argument that the people whose futures are being discussed should be involved in the scenario development process. There is also a pragmatic argument that by involving stakeholders in scenario development, it may be possible to anticipate and prepare for a far wider range of plausible futures than would be the case if scenarios were constructed by researchers alone (Reed *et al.*, 2013).

Stakeholder engagement in scenario development can provide as many benefits for the stakeholders who participate as it does for those who wish to develop the scenarios. Kok *et al.* (2007) and Walz *et al.* (2007) argue that involvement in scenario development may empower participants through the co-generation of new and useful knowledge with researchers, by communicating existing knowledge in a way that can be easily understood, and by increasing participants' capacity to use this knowledge. Stakeholder involvement can provide a wealth of relevant locally held knowledge that might otherwise be missed, and this information may also lead to more pragmatic benefits. In particular, it may be possible to expand the breadth and depth of scenarios, enhancing the logic, internal consistency and validity of the scenarios (Walz *et al.*, 2007). In a context where there was conflict between citizens and planning authorities in Denmark, Tress and Tress (2003) argued that involving stakeholders in scenario development built trust and increased the acceptance of planning decisions by local residents, whilst giving planners access to community knowledge that enabled them to produce better plans. Scenarios and their purposes must nevertheless be fully explained in social and cultural contexts where there are strong beliefs that the future is determined by higher beings and external forces. It also requires careful explanation that scenarios represent merely a range of possible options, not necessarily all of which are desirable. Despite these cautions, scenarios can be useful tools for application at a range of different scales from local to regional and international (Box 7.2).

---

### BOX 7.2: THE SOUTHERN AFRICA MILLENNIUM ECOSYSTEM ASSESSMENT (SAfMA)

The Millennium Ecosystem Assessment (MA) was a 4-year international effort, carried out between 2001 and 2005, to provide decision makers with information on the consequences of ecosystem change for human well-being (MA, 2005). The approach adopted by the MA focused on ecosystem services, including provisioning services such as food and water, as well as regulating services such as flood mitigation, supporting services such as soil formation, and cultural services related to spiritual or aesthetic values. The MA comprised three major components: (i) an assessment of the current condition and trend in the supply of and demand for ecosystem services, (ii) the development of scenarios of plausible future changes in the supply and demand of ecosystem services and the consequences for human well-being and (iii) an analysis of the types of responses that could be implemented to improve ecosystem management and thereby human well-being.

A novel feature of the MA was that it consisted of assessments at the global level as well as at regional and local levels. SAfMA consisted of a cluster of several sub-global assessments undertaken at three different spatial scales in southern Africa (Biggs *et al.* 2004). Five local-scale assessments, each covering the area of a community or local authority, were nested within two basin-scale assessments. The basins were in turn nested within an assessment of the subequatorial continental region. Each SAfMA component study had its own assessment team, consisting of about five dedicated members, and a user advisory group of about ten members who guided the assessors in determining what information users needed. All SAfMA studies assessed three core services: food, water and services linked to biodiversity, as well as additional services of interest to the stakeholders at the particular site or scale.

At the global level, the MA developed four global storylines, which were linked to quantitative models. Initially, SAfMA had considered translating the MA global-scale scenarios down to regional, basin and local levels to develop fully integrated, nested multiscale scenarios. However, it was strongly felt by the SAfMA local-scale groups that such an approach would lead to top-down identification of major uncertainties and thus hold little relevance to their particular situations, in which other factors may be more important. To allow stakeholders the freedom to identify the factors they felt were most important at each scale, and to develop scenarios that were useful and made sense at each particular scale, it was decided that the sub-study groups within SAfMA would construct their scenarios independently.

A diversity of scenario development methods was used in SAfMA. The storylines from four existing regional scenario studies were cross-tabulated against five scenario archetypes derived from the MA (see Scholes and Biggs 2004, Raskin et al. 2005). The key elements of the storylines corresponding to the archetypes Local Resources and Policy Reform, as well as elements from the New Partnership for Africa's Development (NEPAD 2001, 2005), were then synthesized using the MA conceptual framework (MA 2003) to create the two SAfMA regional-scale scenarios. The Gariep Basin assessment used four scenario archetypes, i.e., Fortress World, Market Forces, Local Resources and Policy Reform, and explored their implications for ecosystem services and human well-being in the basin by means of a small expert workshop (Bohensky et al. 2004, Bohensky et al. 2006). The Zambezi Basin assessment based its scenarios on the Intergovernmental Panel on Climate Change Special Report on Emissions Scenarios (IPCC 2002) and used the International Futures Simulation models (Hughes 1999) to quantitatively explore scenarios of poverty, food insecurity, water and energy.

At the local scale, the Gariep local livelihoods assessment identified the key drivers and their likely permutations to derive three qualitative storylines by means of an expert workshop. These scenarios were then converted into theatrical plays and acted out to the local communities to elicit their feedback (Bohensky et al. 2004). In the Gorongosa-Marromeu assessment, two qualitative scenarios were developed in consultation with their user advisory group and presented to the community and other decision makers to explore how they would respond under the different scenarios (Lynam et al. 2004).

Adapted from Kok, K., Biggs, R., and Zurek, M. (2007) Methods for developing multiscale participatory scenarios: insights from southern Africa and Europe. *Ecology and Society* 13(1), 8.

## 7.3 Synthesis

This chapter has considered how best to monitor and evaluate interventions to enhance the capacity of ecosystem and populations to adapt to climate change. The goal is to enable decision-makers to effectively evaluate and then monitor the success of response options so that responses can be improved in the future. In addition to monitoring and evaluating effects of response options on ecosystem processes and services, it is essential to assess the socio-cultural contexts in which adaptations might be implemented, and to evaluate and monitor the effects of those adaptations on livelihoods and human well-being. For this reason, cooperation between members of the policy and research community with practitioners and local communities is important to fully consider the likely implications of different response options and appropriately use monitoring data to refine future responses.

The chapter has considered approaches to monitoring and evaluating current effects of land degradation and climate change. It has considered the benefits and drawbacks of direct measurements, proxy measures (or indicators) and model-based approaches. It concludes that a combination of these approaches is most appropriate for understanding the complex interactions between climate change and land degradation and monitoring their effects. A number of hybrid frameworks and approaches now exist that can enable this combined approach.

Given the complex and uncertain interactions that are likely to take place between land degradation and climate change, it is difficult to predict how different social and ecological systems around the world are likely to be affected by the combined effects of climate change and land degradation. The chapter considered how predictive, visioning and scenario-based approaches may enable policy-makers to better anticipate the likely interactions between land degradation and climate change in future.

# References

Backhaus, R., Bock, M., Weiers, S. 2002. The spatial dimension of landscape sustainability. *Environment, Development and Sustainability* 4: 237–251.

Bajocco, S., De Angelis, A., Perini, L., Ferrara, A., Salvati, L. 2012. The impact of land use/land cover changes on land degradation dynamics: a Mediterranean case study. *Environmental Management* 49: 980–989.

Benayas, J.R., Martins, A., Nicolau, J.M., Schulz, J.J. 2007. Abandonment of agricultural land: an overview of drivers and consequences. *CAB Reviews: Perspectives in Agriculture, Veterinary Science, Nutrition and Natural Resources* 2: 1–14.

Bernoux, M., Chevallier, T. 2014. Carbon in dryland soils. Multiple essential functions. *Les dossiers thématiques du CSFD*. N°10. June 2014. CSFD/Agropolis International: Montpellier, France, 40 pp.

Beskow, S., Avanzi, J.C., Mello, C.R., Viola, M.R, Norton, L.D., Curi, N. 2009. Soil erosion prediction in the grande river basin, Brazil using distributed modeling. *Catena* 79: 49–59.

Biggs, R., Bohensky, E.,. Desanker, P.V, Fabricius, C., Lynam, T., Misslehorn, A. A., Musvoto, C., Mutale, M.,. Reyers, B., Scholes, R. J., Shikongo, S., van Jaarsveld, A. S. 2004. Nature supporting people: the Southern African millennium ecosystem assessment. *Council for Scientific and Industrial Research (CSIR)*. Pretoria, South Africa.

Bohensky, E., Reyers, B., van Jaarsveld, A.S., Fabricius, C. (eds). 2004. *Ecosystem Services in the Gariep Basin*. SUNPReSS: Stellenbosch, South Africa.

Bohensky, E., Reyers, B., van Jaarsveld, A.S. 2006. Future ecosystem services in a southern African river basin: reflections on a scenario planning experience. *Conservation Biology* 20: 1051–1061.

Boix-Fayos, C., Nadeu, E., Quiñonero, J.M., Martínez-Mena, M., Almagro, M. de Vente, J. 2014. Coupling sediment flow-paths with organic carbon dynamics across a Mediterranean catchment. *Hydrology and Earth System Sciences* 11: 5007–5036.

Buenemann, M., Martius, C., Jones, J.W., Herrmann, S.M., Klein, D., Mulligan, M., Reed, M.S., Winslow, M., Washington-Allen, R.A., Ojima., D. 2011. Integrative geospatial approaches for the comprehensive monitoring and assessment of land management sustainability: rationale, potentials, and characteristics. *Land Degradation & Development* 22: 226–239.

Carruthers, G., Tinning, G. 2003. Where, and how, do monitoring and sustainability indicators fit into environmental management systems? *Australian Journal of Experimental Agriculture* 43: 307–323.

Cecchi, G., Mattioli, R.C., Slingenbergh, J., De La Rocque, S. 2008. Land cover and tsetse fly distributions in sub-Saharan Africa. *Medical and Veterinary Entomology* 22: 364–373.

Cerdà, A. 1997. Soil erosion after land abandonment in a semiarid environment of southeastern Spain. *Arid Land Research and Management* 11: 163–176.

Corbiere-Nicollier, T., Ferrari, Y., Jemelin, C., Jolliet, O. 2003. Assessing sustainability: an assessment framework to evaluate Agenda 21 actions at the local level. *International Journal of Sustainable Development and World Ecology* 10: 225–237.

Cramer, V.A., Hobbs, R.J., Standish, R.J. 2008. What's new about old fields? Land abandonment and ecosystem assembly. *Trends in Ecology & Evolution* 23: 104–112.

Dreborg, K.H. 1996. Essence of backcasting. *Futures* 28: 813–828.

Hessel, R., Reed, M.S., Geeson, N., Ritsema, C.J., van Lynden, G., Karavitis, C.A., Schwilch, G., Jetten, V., Burger, P., van der Werff ten Bosch, M.J., Verzandvoort, S., van den Elsen, E., Witsenburg, K. 2014. From framework to action: the DESIRE approach to combat desertification. *Environmental Management* 54: 935–950.

Hughes, B. 1999. *International Futures: Choices in the face of uncertainty.* Westview Press: Boulder, CO, USA.

Innes, J.E., Booher, D.E., 1999. Indicators for sustainable communities: a strategy building on complexity theory and distributed intelligence. *Working Paper 99–04, Institute of Urban and Regional Development.* University of California: Berkeley, CA, USA.

IPCC. 2002. Special report on emissions scenarios (SERS). Available online at: www.grida. no/climate/ipcc/emission/index.htm.

IPCC. 2014. Climate Change 2014. Impacts, Adaptation, and Vulnerability. In: *Part A: global and sectoral aspects. Contribution of Working Group II to the fifth assessment report of the intergovernmental panel on climate change,* Field, C.B., Barros, V. R., Dokken, D. J., Mach, K. J., Mastrandrea, M. D., Bilir, T. E., Chatterjee, M., Ebi, K. L., Estrada, Y.O., Genova, R.C., Girma, B., Kissel, E.S., Levy, A.N., MacCracken, S., Mastrandrea, P. R., White, L. L. (eds). Cambridge University Press: Cambridge, UK and New York, USA.

Kay, A.C. 1989. Predicting the future. *Stanford Engineering* 1: 1–6.

Kok, K., Biggs, R., Zurek, M. 2007. Methods for developing multiscale participatory scenarios: insights from southern Africa and Europe. *Ecology and Society* 13: 8 Available online at: www.ecologyandsociety.org/vol12/iss1/art8/.

Konow, J. 2003. Which is the fairest one of all? A positive analysis of justice theories. *Journal of Economic Literature* 41: 1188–1239.

Lynam, T., Sitoe, A., Reichelt, B., Owen, R., Zolho, R., Cunliffe, R., Bwerinofa, I. 2004. *Human Well-being and Ecosystem Services: An assessment of their linkages in the Gorongosa-Marromeu region of Sofala Province, Mozambique to 2015.* University of Zimbabwe: Harare, Zimbabwe.

MA (Millennium Ecosystem Assessment). 2003. *Ecosystems and Human Well-being: A framework for assessment.* Island Press: Washington, D.C., 245pp.

MA. 2005. *Ecosystems and Human Well-being: Scenarios: Findings of the scenarios working group.* Island Press, Washington, D.C., USA. Available online at: www.maweb.org/en/Scenarios.aspx

Manning, A.D., Lindenmayer, D.B., Fischer, J. 2006. Stretch goals and backcasting: approaches for overcoming barriers to large-scale ecological restoration. *Restoration Ecology* 14: 487–492.

Martin, M.Z., Wullschleger, S.D., Garten, C.T., Palumbo, A.V. 2003. Laser-induced breakdown spectroscopy for the environmental determination of total carbon and nitrogen in soils. *Applied Optics* 42(12): 2072–2077.

Metternicht, G. 2014. Remote sensing of global and regional land degradation processes for improved land governance. *Conference: 35th Asian Conference on Remote Sensing.* At Nay Pyi Taw: Myanmar.

Metternicht, G., Sabelli, A., & Spensley, J. 2014. Climate change vulnerability, impact and adaptation assessment: Lessons from Latin America. *International Journal of Climate Change Strategies and Management* 6(4): 442–476.

NEPAD (New Partnership for Africa's Development). 2001. The new partnership for Africa's development (NEPAD). Available online at: www.nepad.org/2005/files/documents/inbrief.pdf.

NEPAD. 2005. Summary of NEPAD action plans. Available online at: www.nepad.org/2005/files/documents/41.pdf.

Ollinger, S.V., Smith, M.-L. 2005. Net primary production and canopy nitrogen in a temperate forest landscape: an analysis using imaging spectroscopy, modeling and field data. *Ecosystems* 8: 760–778.

Pascual Aguilar, J.A., Sanz García, J., de Bustamante Gutiérrez, I., Kallache, M. 2011. Using environmental metrics to describe the spatial and temporal evolution of landscape structure and soil hydrology and fertility. In: *Proceedings of Spatial 2: Spatial Data Methods for Environmental and Ecological Processes*; Cafarelli, B. (ed). The International Environmetrics Society: Foggia, Italy.

Prell, C., Hubacek, K., Reed, M.S., Burt, T.P., Holden, J., Jin, N., Quinn, C., Sendzimir, J., Termansen, M. 2007. If you have a hammer everything looks like a nail: 'traditional' versus participatory model building. *Interdisciplinary Science Reviews* 32: 1–20.

Raskin, P., Monks, F., Ribeiro, T., van Vuuren, D., Zurek, M. 2005. Global scenarios in historical perspective. In: *Ecosystems and human well being: scenarios. Volume 2: Findings of the scenarios working group of the Millennium Ecosystem Assesment*, Carpenter, S.R., Pingali, P.L., Bennett, E.M., Zurek, M.B. (eds). Island Press, Washington, D.C., USA. Available online at: www.maweb.org/documents/document.326.aspx.pdf.

Reed, M.S., Fraser, E.D.G., Dougill, A.J. 2006. An adaptive learning process for developing and applying sustainability indicators with local communities. *Ecological Economics* 59: 406–418.

Reed, M.S., Buenemann, M., Atlhopheng, J., Akhtar-Schuster, M., Bachmann, F., Bastin, G., Bigas, H., Chanda, R., Dougill, A.J., Essahli, W., Evely, A.C., Fleskens, L., Geeson, N., Glass, J.H., Hessel, R., Holden, J., Ioris, A., Kruger, B., Liniger, H.P., Mphinyane, W., Nainggolan, D., Perkins, J., Raymond, C.M., Ritsema, C.J., Schwilch, G., Sebego, R., Seely, M., Stringer, L.C., Thomas, R., Twomlow, S., Verzandvoort, S. 2011. Cross-scale monitoring and assessment of land degradation and sustainable land management: a methodological framework for knowledge management. *Land Degradation & Development* 22: 261–271.

Reed, M.S., Bonn, A., Broad, K., Burgess, P., Fazey, I.R., Fraser, E.D.G., Hubacek, K., Nainggolan, D., Roberts, P., Quinn, C.H., Stringer, L.C., Thorpe, S., Walton, D.D.,

Ravera, F., Redpath, S. 2013. Participatory scenario development for environmental management: a methodological framework. *Journal of Environmental Management* 128: 345–362.

Riley, J. 2001. Multidisciplinary indicators of impact and change: key issues for identification and summary. *Agriculture, Ecosystems and Environment* 87: 245–259.

Running, S.W., Nemani, R.R., Heinsch, F.A., Zhao, M., Reeves, M., Hashimoto, H. 2004. A continuous satellite-derived measure of global terrestrial primary production. *BioScience* 54: 547–560.

Samuelson, W., Zeckhauser, R. 1988. Status quo bias in decision making. *Journal of Risk and Uncertainty* 1: 7–59.

Scherr, S.J. 1999. Soil degradation: a threat to developing-country food security by 2020? In: *Food, Agriculture, and the Environment Discussion Paper 27*. International Food Policy Research Institute: Washington, D.C., USA.

Scholes, R. J., Biggs, R. 2004. *Ecosystem services in southern Africa: A regional assessment*. Council for Scientific and Industrial Research (CSIR): Pretoria, South Africa.

Stenberg, B., Raphael, A., Rossel, V., Mounem Mouazen, A., Wetterlind, J. 2010. Visible and near infrared spectroscopy in soil science. In: Sparks, D.L. (ed.). *Advances in Agronomy* 107: 163–215.

Stringer, L.C., Reed, M.S., Dougill, A.J., Twyman, C. 2009. Local adaptations to climate change, drought and desertification: insights to enhance policy in southern Africa. *Environmental Science & Policy* 12: 748–765.

Tress, B., Tress, G. 2003. Scenario visualisation for participatory landscape planning a study from Denmark. *Landscape and Urban Planning* 64: 161–178.

UNCCD. 1994. *United Nations Convention to Combat Desertification in Those Countries Experiencing Serious Drought and / or Desertification Particularly in Africa: Text with annexes*. UNEP: Nairobi.

Vanacker, V., Bellin, N., Molina, A., Kubik, P. 2014. Erosion regulation as a function of human disturbances to vegetation cover: a conceptual model. *Landscape Ecology* 29: 293–309.

Walz, A., Lardelli, C., Behrendt, H., Gret-Regamey, A., Lundstrom, C., Kytzia, S., Bebi, P. 2007. Participatory scenario analysis for integrated regional modelling. *Landscape and Urban Planning* 81: 114–131.

WCED (World Commission on Environment and Development). 1987. *Our Common Future*. Oxford University Press: New York.

Wilson, D.C. 1992. Realizing the power of strategic vision. *Long Range Planning* 25: 18–28.

# 8

# MONITORING AND EVALUATING RESPONSE OPTIONS

When evaluating the appropriateness and effectiveness of adaptations to both climate change and land degradation, it is important to consider effects on ecosystem processes and ecosystem services, and how these then impact upon livelihoods, and hence poverty. It is also important to understand the political, institutional, economic and social-technical context in which adaptations may be developed or implemented, to ensure responses are relevant and likely to be adopted and applied effectively (MA, 2005).

The MA (2005) proposed a three-stage assessment process. First it is necessary to identify constraints to adaptation and other response options. Second, trade-offs and synergies associated with different options need to be assessed, evaluating and comparing options in relation to multiple dimensions, focusing on compatibility or conflict between different policy objectives. Third, on this basis it is possible to select adaptation or other response options. The following sections use this framework to consider the role that political and institutional, economic and socio-technical factors are likely to play in creating constraints and trade-offs for adaptation. Methods are then presented for assessing the characteristics of adaptations that may make them more or less likely to be adopted. The chapter concludes by considering methods for monitoring adaptation.

## 8.1 Political and institutional factors

The development and implementation of response options is likely to be significantly hindered if the options face political opposition. It is therefore important to examine the political environment within which responses are being developed. Specifically, it is necessary to consider how adaptation options are likely to be perceived by different political actors (including politicians and stakeholders with political interest) in relation to their agendas. Stakeholder analysis techniques can be used to assess the relative interest and power of these different actors, and can be used to develop engagement strategies to better understand their interests and consider ways of altering adaptations to make them more politically acceptable (Grimble and Chan, 1995, 1997; Reed et al., 2009). It is important however to avoid using such techniques to prioritize the interests of the powerful at the expense of marginalized, disempowered voices within society, particularly when these groups may be most vulnerable to the effects of climate change and land degradation (Reed, 2008; Reed et al., 2009). The extent to which it will be possible to empower these groups in the development of adaptation options will depend on the political structures of the nation state[1] to which they belong, and wider governance structures that shape access to and management of natural resources. These wider governance structures include formal institutions and organizations such as farming co-operatives, as well as informal mechanisms such as behavioural norms and practices within communities of interest (Wenger, 2000).

A range of institutional factors can constrain adaptation options and result in trade-offs or synergies (MA, 2005). Principal among these is capacity for environmental governance. This depends partly on the capacity for individual institutions to govern natural resources, and partly on the interactions between relevant institutions and their collective capacity as a network of actors, underpinned by certain principles, norms and decision-making procedures (Krasner, 1983). Capacity for collaboration at these scales can be constrained by different knowledge bases and understandings of climate change and land degradation, making it challenging to reconcile different perspectives and priorities (Reed et al., 2011). Incentives for collaboration may also be lacking. Where institutions lack skills, information and resources to implement adaptations, the response to climate change and land degradation may fail to protect vulnerable ecosystems and populations. For example, data need to be collected on the implementation of response options, for example via NAPs (UNCCD) and NAPAs (UNFCCC) and associated national reports and communications. Appropriate skills and resources are necessary in order to track progress, and to inform and enable further action.

However, it is possible to build institutional capacity by increasing or pooling access to information, skills and resources. International institutional capacity has grown in recent years through the implementation of, and increasing collaboration between, the Rio Conventions. The effectiveness of international institutions in enabling effective responses to the combined effects of climate change and land degradation depends on the extent to which individual states comply with their commitments under these conventions (Brown-Weiss and Jacobson, 1998; Chasek et al., 2011; Cowie et al., 2011). Compliance with international conventions is more likely if those conventions and the institutions that manage them are perceived to be legitimate and acting within their associated governance mandates (Franck, 1990; Brown-Weiss and Jacobson, 1998). The effectiveness and perceived legitimacy of the Rio Conventions and their associated institutions may be enhanced through more effective collaboration between the Conventions, supported also by adequate resourcing and financing (MA, 2005; Akhtar-Schuster et al., 2011; Requier-Desjardins et al., 2011).

At a national scale, there is a danger that institutional mechanisms addressing climate change and land degradation can become top-down in nature, "being pushed by outside [international] interests" (Dalal-Clayton and Bass, 2009). Top-down institutional structures may not address the needs of those who are actually affected by climate change and land degradation, and may lead to the development of costly solutions that are not adjustable to local contexts and do not effectively reduce vulnerability to climate change or land degradation (Akhtar-Schuster et al., 2011). Political commitment is needed to create an enabling environment for adaptation (Leftwich, 1994). The MA (2005) suggests that to achieve effective adaptation, the national policy environment needs to: be pluralist, enabling multiple interests and ideologies to be represented; and have a clear separation between executive, legislative, and judicial functions, so that decision-making processes can be fully accountable and transparent; and to include environmental policy goals. National institutions are often constrained by a lack of scientifically validated national monitoring and reporting, the results of which are rarely available in politically accessible formats (Akhtar-Schuster et al., 2011). National financial constraints and insufficient legal frameworks and regulations (combined with lack of enforcement) may further limit institutional capacity for adaptation (Dalal-Clayton and Bass, 2009; Akhtar-Schuster et al., 2011).

At a local level, communities have varying capacities for environmental governance, which may enable or hinder adaptation to climate change and land degradation. The capacity for local institutions to enable effective adaptation depends upon:

(i) perceived local benefits from cooperating; (ii) clearly defined rights and boundaries for any natural resources implicated in the response; (iii) knowledge about the state of those resources, including for example their extent, accessibility, and potential for regeneration; (iv) small size of user groups; (v) low degree of heterogeneity of interests and values within user groups; (vi) long-term, multilayered interaction across the communities and other governing institutions involved; (vii) simple, unambiguous rules and adaptable management regimes; (viii) graduated sanctions as punishment; (ix) ease of monitoring and accountability; (x) conflict resolution mechanisms; (xi) strong, effective local leadership; and (xii) congruence with the wider political economy within which those communities function.

*(quoted from MA, 2005)*

Ensuring that the different polycentric institutions and governance levels operate coherently and positively in addressing climate change and land degradation remains an ongoing challenge.

## 8.2 Economic factors

It is increasingly necessary to assess the economic effectiveness of responses on a number of levels, and from a range of perspectives. Often policy and decision makers

---

## BOX 8.1: THE ECONOMICS OF LAND DEGRADATION INITIATIVE

The Economics of Land Degradation Initiative is an international endeavour that brings together stakeholders from the policy, scientific and private sector communities in order to develop relevant data about land and land-based ecosystems. It considers the economic costs of degradation and sustainable land management practices, as well as the costs of inaction in addressing land degradation and desertification. Using valuation and economic approaches, it allows the development of a widely applicable methodology and for common metrics to be produced in order to help policy and decision makers to compare different options, taking into account the full value of land, and reflecting the perspectives of society as a whole.

The ELD Initiative is very timely in the context of both the Sustainable Development Goals (particularly SDG 15: to protect, restore and promote sustainable use of terrestrial ecosystems, sustainably manage forests, combat desertification and halt and reverse land degradation and halt biodiversity loss; UNCSD, 2014) and the global drive towards achieving LDN. If investments are to be made to achieve a land degradation neutral world, investors want to know the magnitude of potential returns.

---

demand the provision of economic figures to help inform their investment priorities. At both national and international scales, it is important to demonstrate the cost-effectiveness of programmes of responses to climate change and land degradation, to justify funding from public sources. The importance of developing methodologies and valuing the land and the ecosystem services it provides has recently advanced through the Economics of Land Degradation Initiative (see Box 8.1).

Fleskens *et al.* (2014) used scenario analysis to evaluate the spatial extent over which potential SLM technologies can be applied, and to calculate their costs and benefits across 13 land degradation hotspots globally. The hotspots are affected by a range of degradation processes including water and wind erosion, drought, wildfires and overgrazing. The study assumed that SLM technologies must be financially attractive in order to be taken up by land managers. Results show that SLM technologies can reduce soil erosion in between 18 percent of study site areas (where vegetative measures are used) to over 50 per cent of study site areas (with the implementation of management measures). Agronomic measures are often cheap. However, average investment costs of technologies vary from €500 per ha for management measures to €1750 per ha for structural and vegetative measures, with important variability both within and between sites. Despite these investment costs, the appraised technologies were financially feasible in 25 per cent (agronomic and management measures) to 100 per cent (vegetative measures) of the areas in which they were found to be applicable. Policy incentives modelled in a policy scenario in many cases led to important gains in the areas where technologies are financially feasible for reducing soil erosion. Yield increases of more than 500 kg per ha are possible in many cases, costing less than €250 per ton of grain calculated over the lifetime of the technologies. Soil erosion can be reduced by 20 to 80 per cent (or more) in the vast majority of cases, generally at a cost of less than €100 per ton of soil conserved. The methodology presented in this study can help to target investment in technologies to the particular degradation hotspots where they will be most cost-effective. There is also scope to assess the viability of technologies that are trialled and implemented in one location in other areas, by updating unit cost price information.

When individuals and local communities consider response options, cost-effectiveness will be evaluated in relation to alternative livelihood strategies, opportunity costs and the financial and other benefits associated with any given option. The value of different adaptations may be considered in a number of different ways, from the "contextual value" of an option in a livelihood context (which may be measured via value indicators including money) to more deeply held "transcendental values". Different response options will be valued differently by different groups within society, depending on the extent to which they are consistent with these more deeply held values.

A range of approaches exist for the valuation of adaptation options, each of which captures some or all of the elements of Total Economic Value[2] (TEV) (IUCN/TNC/World Bank, 2004; Christie et al., 2008). These techniques consider the value of adaptation options in terms of their relative contribution towards different elements of TEV. As such, valuation techniques typically consider the extent to which adaptations may increase an individual's welfare through direct provision of a good (e.g. food, fuel, or recreational use of natural areas), or indirectly through its contribution towards benefits such as the regulation of water and carbon cycles (Pimm et al., 1995; Fromm, 2000). People may also consider the value of adaptations in terms of cultural ecosystem services, e.g. through spiritual or non-use ("passive-use") benefits (such as those derived from cultural values or the knowledge that biodiversity is being protected for future generations to enjoy).

Direct valuation methods typically determine the physical effects of environmental change (such as climate change and land degradation) and measure the monetary value of changes in ecological function and the provision of ecosystem services. Indirect methods assign a monetary value to environmental change, based on production factors affecting the prices of the products (Requier-Desjardins et al., 2011). Some methods are more suited to capturing the value of adaptation options than others. Cost benefit analysis for example, compares the likely impacts of climate change and land degradation with the benefits of adaptation (whether direct or indirect) by translating likely impacts and adaptation benefits into monetary values, to assess whether responses are likely to be cost-effective and generate greater social well-being (Requier-Desjardins et al., 2011). Revealed preference techniques might be more suitable to capturing use values (e.g. the travel cost method which uses the costs of travelling to a biodiversity-rich area to assess the recreation value of that area; Navrud and Mungatana, 1994). On the other hand, stated preference techniques would be more suited to the capture of non-use values (e.g. contingent valuation of how much people are willing to pay to protect an endangered species (Christie, 2007)).

Although environmental valuations have been widely used to assess adaptations to environmental change in both academic and policy-making communities (e.g. Arrow et al., 1993), there has been debate over the validity of these methods (Sagoff, 1988; Diamond and Hausman, 1993; Bate, 1994; Gowdy, 2004). There is also evidence that monetary valuations of ecosystem services are seldom actually used by decision-makers when designing policies, projects and instruments on the ground (Laurans et al., 2013). This lack of uptake in part is linked to a lack of capacity, knowledge and skills to use the results of valuations. There are further concerns about how to adequately capture multiple and complex preferences (Spash, 1993; Spash and Hanley, 1995), especially where intergenerational rights are involved (Bromley, 1995; Hubacek and Mauerhofer, 2008). There has also been recent debate about the way in which values should be elicited. In neoclassical economics, the focus is usually on the expressed preferences of individuals, which are then aggregated and fed into cost-benefit analysis. However, it has also been suggested that people can express preferences as individuals, as individuals in a group setting, or as a group (Clark et al., 2000). Indeed Kenter et al. (2014) found evidence of significant differences between aggregated individual values and values expressed through group deliberative processes. As such, traditional economic analyses can fail to capture the shared, cultural and plural values associated with different adaptation options, given the range of ecosystem services that may be affected by climate change and land degradation. In particular, deeply held cultural values and beliefs that may be shared across a community, and widely divergent preferences that may be placed on the same ecosystem state by different communities (e.g. bush encroachment for cattle versus goat farmers), can easily be omitted (Kenter et al., 2015).

Rather than simply converting the costs and benefits of land degradation and SLM to monetary units, it is important to recognize that people hold different types of values, ranging from attitudinal values or preferences for one type of land management over another, to deeper held ethical or "transcendental values" and beliefs (Kenter et al., 2015). People also play different roles within a single society. This means they may hold different values depending on whether they are asked as an individual land manager or a member of their local community or interest group, or as a consumer versus a citizen. There is also evidence that values around nature are not pre-formed, and are often implicit, and that people may therefore need to form values through deliberation and social interactions with others.

Kenter et al. (2014) show how deliberative processes can inform values, as well as bring out the shared and cultural transcendental values, beliefs and meanings that shape individual values. Deliberative processes also allow participants to consider issues of fairness, risk and uncertainty more explicitly, and take into account the medium- and long-term impacts of a decision. For this reason, mixed method and participatory approaches to environmental valuation are growing in popularity, combining monetary valuation tools (such as Cost-Benefit Analysis and Deliberative Monetary Valuation) with non-monetary valuation tools (such as Multi-Criteria Evaluation or Matrix Ranking) (Wegner and Pascual 2011; Parks and Gowdy 2013; Kenter et al., 2014). In this way, it is argued that it may be possible to more robustly and fairly assess the social impacts of adaptation options, as well as their economic impacts, informed by locally-held knowledge and the context in which climate change and land degradation occurs (Farber et al., 2002; Bebbington et al., 2007; Fujiwara and Campbell 2011; Parks and Gowdy 2013). Good practices in stakeholder participation and deliberation are considered in the next chapter.

## 8.3 Social-technical factors

Many early adaptation studies assumed that adaptation was primarily a function of available technology and technical knowledge (Burton et al., 2002; van Aalst et al., 2008). However, these top-down approaches failed to consider local socio-technical constraints to the development and implementation of adaptations (e.g. access, cost and the necessary skills), or the influence of local socio-cultural contexts on adaptation choices (van Aalst et al., 2008). Innovation may be particularly important in the development of adaptation options that can simultaneously enable adaptation to climate change and land degradation. Innovation in this context means "an idea, practice, or object that is perceived as new by an individual or other unit of adoption" (Rogers 1995: 11).

A large body of work exists to evaluate, refine and disseminate innovative adaptation options. Much of this work has focused on agricultural innovations and soil and water conservation. Rogers (1995) describes adoption as a five step "innovation-decision process" in which farmers: (i) gain knowledge of an innovation; (ii) seek information about the likely consequences of adoption and form an attitude towards it; (iii) decide to adopt or reject the innovation; (iv) implement the innovation; and (v) confirm their innovation decision by seeking reinforcement, and discontinue it if exposed to conflicting experiences and messages. Rogers (1995) also identified five key perceived characteristics of innovations that determine their adoption potential: relative advantage, trialability, compatibility, observability and complexity. The most significant of these for adoption are usually high relative advantage, high compatibility and low complexity (Tornatzky and Klein, 1982). Reed (2007) added adjustability: the extent to which an innovation can be adjusted to meet dynamic, and sometimes unforeseen, user demands and specifications. Furthermore, Reed (2007) integrated the innovation-decision process with the sustainable livelihoods framework, suggesting that the need to innovate was stimulated by farmer needs and aspirations, which in turn were influenced by their changing endowment and access to capital assets. At the same time, the perceived risk associated with an innovation is negatively related to its rate of adoption. Perceived risk is the degree to which economic, social, physical and functional risks are perceived as being associated with the innovation (Slovic, 1987). Risk perception is influenced by the interaction of individual psychological, social and other cultural factors, and the subsequent behavior of individuals and groups may further affect the way these risks are perceived (Kasperson *et al.*, 1988; Kasperson and Kasperson, 2005). These sorts of approaches stand in contrast to traditional economic approaches, which tend to assume that people have complete knowledge of the system within which they are adapting, and apply this knowledge through economically rational behaviour and decision making to optimize profits (Ellis, 1988; Parker *et al.*, 2008). However, diffusion theory has been criticized for being used as a highly structured and top-down tool that tends to be used by the powerful to influence others. It also assumes that well-connected social networks exist through which innovations can diffuse, which is not always the case (Reed *et al.*, 2013a).

Partly in reaction to this, there is now growing interest in the role that social learning might play in developing and diffusing adaptations to climate change. Reed *et al.* (2010) argue that to be considered "social learning", a process must: (i) demonstrate that some depth of conceptual change or change in understanding has taken place in the individuals involved; (ii) demonstrate some degree of breadth for this change to go beyond individuals and become situated within wider social groups within society; and (iii) occur through social interactions and processes between actors within a social network. Such learning is typically accompanied by individual and group reflection about adaptations, and iterative attempts to apply what is learned, making incremental changes to the socio-ecological system (Forester 1999; Daniels and Walker 2001; Schusler and Decker 2003; Keen and Mahanty 2006). Pelling *et al.* (2008) argue that adaptive behaviour is by definition a form of learning. As such, they argue that it is essential to understand the processes through which people learn how to be adaptive. Drawing on social learning theory, they propose that to develop innovative adaptation options and permit their wider diffusion, it is necessary for institutions to create "spaces" in which individuals and groups can experiment, communicate, learn and reflect on new ideas. It should be noted that as social learning takes place through interaction within social networks, the network characteristics can hinder or promote the development and dissemination of adaptation options (Pelling and High, 2005). For example, social networks may rapidly diffuse effective and socially acceptable adaptations but social norms or traditional taboos may prevent the adoption of other adaptations (Reed *et al.*, 2013a).

## 8.4 Monitoring adaptation

Despite an extensive literature on monitoring climate change and land degradation processes and impacts, attention has only recently moved to monitoring adaptations to these processes (Carpenter *et al.*, 2006; Schwilch *et al.*, 2011). Although there have been many assessments highlighting successful adaptation to climate change and land degradation (e.g. by UNEP, 2002; Pretty and Koohafkan, 2002; GM-CCD – Reij and Steeds, 2003; IWMI – Penning de Vries *et al.*, 2008; and UNFCCC, 2010) these have typically been cross-sectional snapshots in time, and have seldom involved long-term monitoring (Hedger *et al.*, 2008; Prowse and Sniltveit, 2010; Schwilch *et al.*, 2011). Ultimately, the goal of monitoring is to provide information about the effects of adaptation and other responses to climate change and land degradation on livelihoods and human well-being. Given the mechanisms through which climate change and land degradation are likely to interact with one another, this is likely to involve monitoring changes in ecosystem processes and the provision of ecosystem services, as well as socio-economic assessments that can demonstrate links between adaptation, livelihoods and well-being. When adaptation monitoring does happen, it is typically done (informally) by land managers, or by external experts (often extension services). However, there are few documented methods for monitoring adaptation responses in the academic literature (Schwilch *et al.*, 2011).

Although not designed to monitor the effectiveness of adaptation over time, land capability assessments (Helms, 1992) have formed the basis for some adaptation assessments, given their ability to assess changes in the productive potential of the land in relation to soil quality, land use and climate. For example, land capability models have been used to assess the likely productivity of agricultural land under climate change scenarios and hence the types of land uses and crops that may be grown in future (Brown *et al.*, 2008, 2011). In some cases, these models have been linked to soil erosion models to consider how such adaptations might interact with land degradation processes (e.g. Reed *et al.*, 2013b). As such, with appropriate data sources, it may be possible to use techniques based on land capability assessment to monitor the extent to which adaptations to climate change and land degradation maintain the productivity of land for agriculture, and hence in theory support livelihoods.

Where it is possible to monitor the effects of climate change and land degradation via remote sensing, it may be possible to monitor the extent to which response options mitigate or reverse these effects (Buenemann *et al.*, 2011). Remote sensing has been used to monitor the provision of many ecosystem services, including: provisioning services (e.g. land cover, biological productivity, and water quality and quantity via products such as GlobCover and Afri-Cover); regulating services (e.g. climate variables such as temperature, diseases and evapotranspiration); supporting services (e.g. nutrient cycling and soil characteristics such as soil organic matter, soil moisture, surface roughness, texture, moisture and salinity) (Buenemann *et al.*, 2011); and cultural services (e.g. "wild" land mapping; Carver *et al.*, 2012, 2013). However, determining causality would normally require more detailed, field-level biophysical and socio-economic assessments, given the wide range of other factors that could account for any changes.

More local-scale methods for monitoring adaptation have been developed recently by the LADA[3] and DESIRE[4] projects in collaboration with WOCAT[5]. Although these projects focussed primarily on monitoring SLM practices, SLM is a key adaptation to both climate change and land degradation and the approach could be, in principle, easily extended to monitor the biophysical and socio-economic effects of other response options. The WOCAT methodology at the local level involves:

1.  assessing local case studies of successful response options and their local spread and adoption;
2.  using a standardized framework that enables comparison and transferability beyond the local scale;
3.  inclusion of socio-economic as well as biophysical aspects;
4.  use of the knowledge of both specialists and land users as data sources, backed up (triangulated) by scientific data where possible; and
5.  simultaneously using the same tools for both (self-) evaluation and for knowledge sharing (based on Schwilch et al., 2011).

This is done at the local level using two questionnaires, on specific SLM technologies (physical practices and management measures that control land degradation and enhance productivity) and broader approaches to SLM (which enable the implementation, adjustment and uptake of these technologies), which feed into an online database. Importantly, the WOCAT tools enable self-evaluation by land managers, which can inform and refine their practices, as well as being applicable by external assessors.

The use of standardized methods and indicators creates the possibility of comparing progress between locations internationally, and integrating information about the success of response options at multiple scales, from local to international (Schwilch et al., 2011). Reed et al. (2011) suggest that a nested approach in which indicators would be developed at local levels, relevant to monitoring adaptations in different contexts, but with a core set of indicators monitored across contexts to enable cross-scale comparisons and global-scale monitoring. Linked to this the UN-INWEH's KM:Land[6] project developed an online Learning Network for SLM practitioners to accompany an indicator system for monitoring the impact of SLM projects (funded by the GEF) that can compare progress between projects internationally.

Adaptation indicators may be process-based (to help measure progress and enhance adaptation) or outcome-based (measuring the effectiveness of interventions). The characteristics of effective indicators are universal, and can be broadly grouped under criteria relating to their robustness (i.e. they should be accurate, free from bias, reliable and consistent over space and time, assess trends over time, provide early warnings, be verifiable and replicable, provide timely information and be measured in relation to relevant targets or baselines) and ease of use (i.e. they should be easily measured, make use of available data, have social appeal and resonance with users, be cost-effective and rapid to use, provide clear and unambiguous information, be limited in number and be easily accessible to decision-makers) (Reed et al., 2006). Effective indicators for assessing adaptation need to consider: effectiveness (the extent to which objectives are achieved); flexibility (to account for the uncertainty of climate change and land degradation processes, and the evolving knowledge base); equity (across sectors, regions, scales and societies); efficiency (to address agreed acceptable levels of risk); and sustainability of adaptation interventions (including partnership-building and community engagement) (Hedger *et al.*, 2008). Indicators may be developed to cover thematic areas (e.g. by sector, habitat or issue), and may be developed to represent adaptation processes such as policy and planning processes, capacity development and awareness raising.

Developing indicators at local or project scales is relatively straightforward, and can often use existing indicators and datasets from established monitoring systems. Indeed, Spearman and McGray (2011) argue that indicators should normally be developed during project design and linked to project objectives in order to ensure progress can be monitored. They go on to propose a six-step process for developing a monitoring and assessment system for an adaptation project:

1. *Describe the adaptation context*: conducting a climate change and land degradation vulnerability or risk assessment early in the design process can help practitioners to understand how ecosystems and human populations are likely to be affected by climate change and land degradation, and to identify which aspects the adaptation intervention might be able to address;
2. *Identify the contribution to adaptation*: identify ways an intervention can contribute towards adaptive capacity, specific adaptation actions and sustained development;
3. *Form an adaptation hypothesis*: to evaluate the validity of specific approaches to adaptation, practitioners can formulate an adaptation hypothesis for each of the expected outcomes of their intervention;
4. *Create an adaptation "theory of change"*: to illustrate the relationship between an intervention's components, expected results and assumptions about factors that might affect its likelihood of achieving success;
5. *Choose indicators and set a baseline*: indicators may be characterized by type of outcome and should link to a baseline from which progress can be measured; and
6. *Use the adaptation monitoring and evaluation system*: its implementation needs to involve flexibility and learning, via regular feedback and engagement with partners.

In any case, monitoring and assessment of adaptation responses should be carried out through robust Impact Evaluation (IE) methods which scientifically measure the causal effect of a specific adaptation intervention *vis-à-vis* a credible counterfactual scenario and seek to understand the conditions under which effects arise. In such a robust IE, building on the adaptation "theory of change", there is a need to rule out rival explanations of the outcomes of the programme (adaptation response and associated designed instrument, e.g. a payment scheme for environmental services) so that evaluators can attribute impacts specifically to the programme. These types of IE, including quasi-experimental methods and randomized controlled trials, have been developed in the health, education and development sectors and more recently in the field of conservation (Ferraro and Hanauer, 2014). IE could be applied in the field of responses to vulnerability to land degradation and climate change. In turn, this would efficiently inform policy-makers about what works and what does not.

However, monitoring of adaptation policies and national, cross-sectoral adaptation programmes is more complex. In particular, there are challenges in establishing cause and effect in adaptation interventions and accounting for unintended consequences. There are rarely baselines or historical trends, from which progress may be inferred, and more generally there tends to be limited sharing of relevant information across stakeholder groups, levels and sectors (Hedger *et al.*, 2008; Prowse and Sniltveit, 2010). Nevertheless, Frankel-Reed *et al.* (2009: 285) suggest that monitoring and evaluation of adaptation at these broader scales is possible, and should:

> identify links between development and climate change; focus the scope of adaptation on key sectors, themes or issues (without limiting integration among them); identify processes, institutions and capacities to strengthen system-wide adaptive capacity; identify adaptation practices/behaviours related to development outcomes (including the actors involved and the change sought); identify adaptation measures necessary to reduce climate-related [and land degradation] risks; incorporate climate hazard and capacity/vulnerability factors; and balance quantitative, qualitative and narrative monitoring and evaluation tools to enable triangulation.

Some general lessons can however be derived for monitoring adaptation to climate change and land degradation. Given the nature of climate change and land degradation processes and the types of feedbacks that may occur between these two processes, monitoring and evaluation needs to consider both biophysical and socio-economic changes arising from adaptations. There are a number of biophysical indicators that may be monitored cost-effectively via remote-sensing at broad spatial scales. However, field-based measurements are likely to be necessary to interpret this data, and to establish cause and effect. Although field-based measurements of biophysical indicators are time-consuming and expensive, it may in some cases be possible to develop indicators that can be used by land managers to inform their adaptation practice, which can also be reported and used at broader spatial scales. By taking a nested approach to indicator development, it may be possible to develop locally relevant indicators around a core set of indicators that can enable cross-scale comparisons. Even with more detailed field-based data, it may be difficult to directly attribute changes to adaptation interventions. Socio-economic (often qualitative) data is also essential to triangulate and supplement biophysical data, in order to understand whether observed changes in biophysical variables (e.g. increase vegetation cover and biomass) may be considered to be sustainable (e.g. if the vegetation is palatable to livestock) or whether land degradation is worsening further (e.g. if the vegetation represents encroachment by unpalatable species).

Such data is also necessary to understand changes in natural capital (which may be observed using biophysical indicators) in the context of changes in other capital assets (e.g. financial, physical, human or social capital), to interpret the overall impact of interventions on livelihoods and well-being. Finally, if the goal of adaptation is to promote sustainable livelihoods and human well-being, then there is a strong normative argument for engaging affected communities in monitoring, so that adaptations may be refined using local knowledge and made more relevant to local needs and priorities (Abbot and Guijt, 1997; Reed *et al.*, 2006).

## 8.5 Synthesis

The chapter has considered how responses to the interactive effects of climate change and land degradation may be assessed, considering the political, institutional, economic and social-technical context in which adaptations may be developed or implemented, to ensure responses are relevant and likely to be adopted and applied effectively. It then reviewed methods for monitoring the success of adaptation interventions.

Given types of interactions likely to occur between climate change and land degradation in future, monitoring and evaluation needs to consider both biophysical and socio-economic changes arising from adaptations. There are a number of biophysical indicators that may be monitored cost-effectively via remote-sensing at broad spatial scales. However, field-based measurements are likely to be necessary to interpret this data, and to establish cause and effect. Even with more detailed field-based data, it may be difficult to directly attribute changes to adaptation interventions. Socio-economic (often qualitative) data is therefore essential to triangulate and supplement biophysical data, in order to understand whether observed changes in biophysical variables may be considered to be sustainable or are further worsening land degradation. Such data is also necessary to understand changes in natural capital in the context of changes in other capital assets, to interpret the overall impact of interventions on livelihoods and well-being.

## Notes

1 MA (2005: 76) define a nation state as "a combination of actors and institutions, encompassing manifold activities that include everything from political fundraisers, legislative committee hearings, and consultative meetings, to policy implementation on the ground".
2 This comprises both use values that accrue directly or indirectly to those who use (or wish to have the option to use) environmental resources (Bateman *et al.*, 2002) and non-use values that reflect the value individuals attach to environmental resources even if they do not use them, for example the value of simply knowing that these resources exist or the value of preserving them for future generations (Krutilla, 1967; Smith, 1987 and Cameron, 1992).
3 http://www.fao.org/nr/lada/
4 http://www.desire-project.eu
5 https://www.wocat.net
6 http://www.comap.ca/kmland/

# References

Abbot, J., Guijt, I. 1997. *Changing views on change: A working paper on participatory monitoring of the environment.* Working Paper: International Institute for Environment and Development, London, 112 pp.

Akhtar-Schuster, M., Thomas, R. J., Stringer, L.C., Chasek, P., Seely, M K. 2011. Improving the enabling environment to combat land degradation: institutional, financial, legal and science-policy challenges and solutions. *Land Degradation and Development* 22: 299–312.

Arrow, K., Solow, R., Portney, P., Leamer, E., Radner, R., Schuman, H. 1993. Report of the NOAA panel on contingent valuation. *Federal Register* 58: 4602–4614.

Bate, R. 1994. *Pick a Number: A critique of contingent valuation methodology and its application in public policy.* Competitive Enterprise Institute: Washington, D.C., USA.

Bateman, I., Carson, R., Day, B., Hanemann, W.M., Hanley, N., Hett, T., Jones-Lee, M., Loomes, G., Mourato, S., Ozdemiroglu, E., Pearce, D.W., Sugden, R., Swanson. S. 2002. *Guidelines for the Use of Expressed Preference Methods for the Valuation of Preferences for Non-market Goods.* Edward Elgar: Cheltenham.

Bebbington, J., Brown, J., Frame, B. 2007. Accounting technologies and sustainability assessment models. *Ecological Economics* 61: 224–236.

Bromley, D.W. (ed.). 1995. *The Handbook of Environmental Economics.* Oxford: Blackwell.

Brown, I., Towers, W., Rivington, M., Black, H.I.J. 2008. The influence of climate change on agricultural land-use potential: adapting and updating the land capability system for Scotland. *Climate Research* 37: 43–57.

Brown, I., Poggio, L., Gimona, A., Castellazzi, M.S. 2011. Climate change, drought risk and land capability for agriculture: implications for land-use in Scotland. *Regional Environmental Change* 11: 503–518.

Brown-Weiss, E., Jacobson, H. (eds), 1998. *Engaging Countries: Strengthening compliance with international environmental accords.* MIT Press: Cambridge, MA.

Buenemann, M., Martius, C., Jones, J.W., Herrmann, S.M., Klein, D., Mulligan, M., Reed, M.S., Winslow, M., Washington-Allen, R.A., Ojima, D. 2011. Integrative geospatial approaches for the comprehensive monitoring and assessment of land management sustainability: rationale, potentials, and characteristics. *Land Degradation & Development* 22: 226–239.

Burton, I., Huq, S., Lim, B., Pilifosova, O., Schipper, E.L. 2002. From impacts assessment to adaptation priorities: the shaping of adaptation policy. *Climate Policy* 2: 145–159.

Cameron, T. A. 1992. Nonuser resource values. *American Journal of Agricultural Economics* 74: 1133–1137.

Carpenter, S.R., DeFries, R., Dietz, T., Mooney, H.A., Polasky, S., Reid, W.V., Scholes, R.J. 2006. Millennium ecosystem assessment: research needs. *Science* 314: 257–258.

Carver, S., Nutter, S., Comber, A., McMorran, R. 2012. A GIS model for mapping spatial patterns and distribution of wild land in Scotland. *Landscape and Urban Planning* 104: 395–409.

Carver, S., Tricker, J., Landres, P. 2013. Keeping it wild: mapping wilderness character in the United States. *Journal of Environmental Management* 131: 239–255.

Chasek, P., Essahli, W., Akhtar-Schuster, M., Stringer, L.C., Thomas, R. 2011. Integrated land degradation monitoring and assessment: horizontal knowledge management at the national and international levels. *Land Degradation & Development* 22: 272–284.

Christie, M., 2007. An examination of the disparity between hypothetical and actual willingness to pay for Red Kite conservation using the contingent valuation method. *Canadian Journal of Agricultural Economics* 55: 159–169.

Christie, M., Fazey, I., Cooper, R., Hyde, T., Deri, A., Hughes, L., Bush, G., Brander, L., Nahman, A., de Lange, W., Reyers, B. 2008. *An Evaluation of Economic and Non-economic*

*Techniques for Assessing the Importance of Biodiversity to People in Developing Countries*. Defra: London.

Clark, J., Burgess, J., Harrison, C. 2000. "I struggled with this money business": respondents' perspectives on contingent valuation. *Ecological Economics* 33: 45–62.

Cowie, A.L., Penman, T.D., Gorissen, L., Winslow, M.D., Lehmann, J., Tyrrell, T.D., Twomlow, S., Wilkes, A., Lal, R., Jones, J.W., Paulsch, A., Kellner, K., Akhtar-Schuster, M. 2011. Towards sustainable land management in the drylands: scientific connections in monitoring and assessing dryland degradation, climate change and biodiversity. *Land Degradation & Development* 22: 248–260.

Dalal-Clayton, B., Bass, S. 2009. The challenges of environmental mainstreaming: experiences of integrating environment into development institutions and decisions. *IIED*. Available online at: www.iied.org/pubs/pdfs/17504IIED.pdf.

Daniels, S.E., Walker, G.B., 2001. *Working Through Environmental Conflict: The collaborative learning approach*. Praeger: Westport, CT, USA.

Diamond, P.A., Hausman, J.A. 1993. On contingent valuation measurement of nonuse values. In: Hausman, J.A. (ed.). *Contingent Valuation. A Critical Assessment*. North-Holland: Amsterdam, pp. 3–38.

Ellis, F.M. 1988. *Peasant Economics*. Cambridge University Press: Cambridge.

Farber, S.C., Costanza, R., Wilson, M.A. 2002. Economic and ecological concepts for valuing ecosystem services. *Ecological Economics* 41: 375–392.

Ferraro, P.J., Hanauer, M.M. 2014. Advances in measuring the environmental and social impacts of environmental programs. *Annual Review of Environment and Resources* 39: 495–517.

Fleskens, L., Nainggolan, D., Stringer, L.C. 2014. An exploration of scenarios to support sustainable land management using integrated environmental socio-economic models. *Environmental Management* 54: 1005–1021.

Forester, J. 1999. *The Deliberative Practitioner*. MIT Press: Cambridge, MA.

Franck, T.M. 1990. *The Power of Legitimacy among Nations*. Oxford University Press: Oxford, UK.

Frankel-Reed, J., Brooks, N., Kurukulasuriya, P., Lim, B. 2009. A framework for evaluating adaptation to climate change. In: *Evaluating Climate Change and Development*, Van den Berg, R. D., Feinstein, O.N. (eds). *World Bank Series on Development*, Volume 8. Transaction Publishers. pp. 285–298.

Fromm, O. 2000. Ecological structure and functions of biodiversity as elements of its total economic value. *Environmental and Resource Economics* 16: 303–328.

Fujiwara, D., Campbell, R. 2011. *Valuation Techniques for Social Cost-benefit Analysis – Stated Preference, Revealed Preference and Subjective Well-Being Approaches, a Discussion of the Current Issues*. HM Treasury, London.

Gowdy, J.M. 2004. The revolution in welfare economics and its implications for environmental valuation and policy. *Land Economics* 80: 239–257.

Grimble, R., Chan, M.K. 1995. Stakeholder analysis for natural resource management in developing countries. *Natural Resources Forum* 19: 113–124.

Grimble, R., Wellard, K. 1997. Stakeholder methodologies in natural resource management: a review of principles, contexts, experiences and opportunities. *Agricultural Systems* 55: 173–193.

Hedger, M., Horrocks, L., Mitchell, T., Leavy, J. Greeley, M. 2008. *Evaluation of Adaptation to Climate Change from a Development Perspective*. Desk Review, Institute of Development Studies: Brighton.

Helms, D. 1992. The development of the land capability classification. Soil Conservation Service: Washington, D.C., USA. Available online at: www.nrcs.usda.gov/about/history/articles/LandClassification.html

Hubacek, K., Mauerhofer, V. 2008. Future generations: economic, legal and institutional aspects. *Futures* 40: 413–423.

IUCN, TNC and the World Bank. 2004. *How Much is an Ecosystem Worth? Assessing the economic value of conservation*. IUCN/TNC/IBRD/The World Bank, Washington, D.C., USA.

Kasperson, R.E., Renn, O., Slovic, P., Brown, H., Emel, J., Goble, R., Kasperson, J., Ratick, J. 1988. The social amplification of risk: a conceptual framework. *Risk Analysis* 8: 177–187.

Kasperson, J.X., Kasperson, R.E. 2005. *The Social Contours of Risk. Volume I: Publics, risk communication and the social amplification of risk*. Earthscan: Virginia, USA.

Keen, M., Mahanty, S. 2006. Learning in sustainable natural resource management: challenges and opportunities in the Pacific. *Society and Natural Resources* 19: 497–513.

Kenter, J.O., Reed, M.S., Irvine, K.N., O'Brien, E., Brady, E., Bryce, R., Christie, M., Church, A., Cooper, N., Davies, A., Hockley, N., Fazey, I., Jobstvogt, N., Molloy, C., Orchard-Webb, J., Ravenscroft, N., Ryan, M., Watson, V. 2014. *UK National Ecosystem Assessment Follow-on Phase, Technical Report: Shared, plural and cultural values of ecosystems*. UNEP-WCMC: Cambridge.

Kenter, J.O., Brady, E., Bryce, R., Christie, M., Church, A., Irvine, K.N., Cooper, N., Davies, A., Evely, A., Everard, M., Fazey, I., Hockley, N., Jobstvogt, N., Molloy, C., O'Brien, L., Orchard-Webb, J., Ravenscroft, N., Ranger, S., Reed, M.S., Ryan, M., Watson, V. 2015. What are shared and social values of ecosystems? *Ecological Economics* 111: 86–99.

Krasner S.D. 1983. Structural causes and regime consequences: regimes as intervening variables. In: *International Regimes*; Krasner, S.D. (ed.). Cornell University Press: Ithaca, pp. 1–22.

Krutilla, J.V. 1967. Conservation reconsidered. *The American Economic Review* 57, 777–786.

Laurans, Y., Rankovic, A., Billé, R., Pirard, R., Mermet, L. 2013. Use of ecosystem services economic valuation for decision making: questioning a literature blindspot. *Journal of Environmental Management* 119: 208–219.

Leftwich, A. 1994. Governance, the state and the politics of development. *Development and Change* 25: 363–386.

MA (Millennium Ecosystem Assessment). 2005. *Ecosystems and Human Well-being: Policy Responses*. Island Press, Washington, D.C., USA.

Navrud, S., Mungatana, E.D. 1994. Environmental valuation in developing countries: the recreational value of wildlife viewing. *Ecological Economics* 11: 135–151.

Parker, D.C., Hessl, A., Davis, S.C. 2008. Complexity, land-use modeling, and the human dimension: fundamental challenges for mapping unknown outcome spaces. *Geoforum* 39: 789–804.

Parks, S., Gowdy, J. 2013. What have economists learned about valuing nature? A review essay. *Ecosystem Services* 3: e1–e10.

Pelling, M., High, C. 2005. Understanding adaptation: what can social capital offer assessments of adaptive capacity? *Global Environmental Change* 15: 308–319.

Pelling, M., High, C., Dearing, J., Smith, D. 2008. Shadow spaces for social learning: a relational understanding of adaptive capacity to climate change within organisations. *Environment and Planning A* 40: 867–884.

Penning de Vries, F., Acquay, H., Molden, D., Scherr, S., Valentin, C., Olufunke, C. 2008. Learning from bright spots to enhance food security and to combat degradation of water and land resources. In: *Conserving Land, Protecting Water, Comprehensive Assessment of Water Management in Agriculture Series, Volume 6*, Bossio, D., Geheb, K., (eds). CABI: Wallingford, pp. 1–19.

Pimm, S.G.R., Gittleman, J., Brooks, T. 1995. The future of biodiversity. *Science* 269: 247–350.

Pretty, J., Koohafkan, P. 2002. *Land and Agriculture: A compendium of recent sustainable development initiatives in the field of agriculture and land management.* Food and Agriculture Organization. Rome, 59 pp.

Prowse, M., Snilstveit, B. 2010. *Impact Evaluation and Interventions to Address Climate Change: A scoping study.* The International Initiative for Impact Evaluation (3ie): New Dehli.

Reed, M.S., Fraser, E.D.G., Dougill, A.J. 2006. An adaptive learning process for developing and applying sustainability indicators with local communities. *Ecological Economics* 59: 406–418.

Reed, M.S. 2007. Participatory technology development for agroforestry extension: an innovation-decision approach. *African Journal of Agricultural Research* 2: 334–341.

Reed, M.S. 2008. Stakeholder participation for environmental management: a literature review. *Biological Conservation* 141: 2417–2431.

Reed, M.S., Graves, A., Dandy, N., Posthumus, H., Hubacek, K., Morris, J., Prell, C., Quinn, C.H., Stringer, L.C. 2009. Who's in and why? Stakeholder analysis as a prerequisite for sustainable natural resource management. *Journal of Environmental Management* 90: 1933–1949.

Reed, M.S., Evely, A.C., Cundill, G., Fazey, I., Glass, J., Laing, A., Newig, J., Parrish, B., Prell, C., Raymond, C., Stringer, L.C. 2010. What is social learning? *Ecology & Society* 15: r1. Available online at: www.ecologyandsociety.org/vol15/iss4/resp1/.

Reed, M.S., Buenemann, M., Atlhopheng, J., Akhtar-Schuster, M., Bachmann, F., Bastin, G., Bigas, H., Chanda, R., Dougill, A.J., Essahli, W., Evely, A.C., Fleskens, L., Geeson, N., Glass, J.H., Hessel, R., Holden, J., Ioris, A., Kruger, B., Liniger, H.P., Mphinyane, W., Nainggolan, D., Perkins, J., Raymond, C.M., Ritsema, C.J., Schwilch, G., Sebego, R., Seely, M., Stringer, L.C., Thomas, R., Twomlow, S., Verzandvoort, S. 2011. Cross-scale monitoring and assessment of land degradation and sustainable land management: a methodological framework for knowledge management. *Land Degradation & Development* 22: 261–271.

Reed, M.S., Podesta, G., Fazey, I., Beharry, N.C., Coen, R., Geeson, N., Hessel, R., Hubacek, K., Letson, D., Nainggolan, D., Prell, C., Psarra, D., Rickenbach, M.G., Schwilch, G., Stringer, L.C., Thomas, A.D. 2013a. Combining analytical frameworks to assess livelihood vulnerability to climate change and analyse adaptation options. *Ecological Economics* 94: 66–77.

Reed, M.S., Hubacek, K., Bonn, A., Burt, T.P., Holden, J., Stringer, L.C., Beharry-Borg, N., Buckmaster, S., Chapman, D., Chapman, P., Clay, G.D., Cornell, S., Dougill, A.J., Evely, A., Fraser, E.D.G., Jin, N., Irvine, B., Kirkby, M., Kunin, W., Prell, C., Quinn, C.H., Slee, W., Stagl, S., Termansen, M., Thorp, S., Worrall, F. 2013b. Anticipating and managing future trade-offs and complementarities between ecosystem services. *Ecology & Society* 18: 5.

Reij, C., Steeds, D. 2003. Success stories in Africa's drylands: supporting advocates and answering critics. *Global Mechanism of the Convention to Combat Desertification (GM-CCD).* CIS/Centre for International Cooperation Vrije Universiteit Amsterdam, 32 pp.

Requier-Desjardins, M., Adhikari, B., Sperlich, S. 2011. Some notes on the economic assessment of land degradation. *Land Degradation & Development*, 22: 285–298.

Rogers, E.M. 1995. *Diffusion of Innovations*, 4th Edition. The Free Press: New York.

Sagoff, M. 1988. *The Economy of the Earth: Philosophy, law and the environment.* Cambridge University Press: Cambridge.

Schusler, T.M., Decker, D.J. 2003. Social learning for collaborative natural resource management. *Society and Natural Resources* 15: 309–326.

Schwilch, G., Bestelmeyer, B., Bunning, S., Critchley, W., Herrick, J., Kellner, K., Liniger, H., Nachtergaele, F., Ritsema, C., Schuster, B., Tabo, R., van Lynden, G., Winslow, M. 2011. Experiences in monitoring and assessment of sustainable land management. *Land Degradation & Development* 22: 214–225.

Slovic, P. 1987. Perception of risk. *Science* 236: 280–285.

Smith, V.K. 1987. Non-use values in benefit cost analysis. *Southern Economic Journal* 54: 19–26.

Spash, C.L. 1993. Economics, ethics and long term environmental damages. *Environmental Ethics* 15: 117–132.

Spash C.L., Hanley N. 1995. Preferences, information and biodiversity preservation. *Ecological Economics* 12: 191–208.

Spearman, M., McGray, H. 2011. Making adaptation count: concepts and options for monitoring and evaluation of climate change adaptation. *Deutsche Gesellschaft für Internationale Zusammenarbeit (GIZ) GmbH*: Eschborn, Germany.

Tornatzky, L.G., Klein, K.J. 1982. Innovation characteristics and innovation adoption-implementation: a meta-analysis of findings. *IEEE Transactions on Engineering Management* 29: 28–45.

UNCSD. 2014. Open working group proposal for sustainable development goals. Available online at: https://sustainabledevelopment.un.org/focussdgs.html.

UNEP. 2002. *Success Stories in the Struggle Against Desertification.* UNEP: Nairobi.

UNFCCC. 2010. Synthesis report on efforts undertaken to monitor and evaluate the implementation of adaptation projects, policies and programmes and the costs and effectiveness of completed projects, policies and programmes, and views on lessons learned, good practices, gaps and needs. UNFCCC: Bonn.

Van Aalst, M.K., Cannon, T., Burton, I. 2008. Community level adaptation to climate change: the potential role of participatory community risk assessment. *Global Environmental Change* 18: 165–179.

Wenger, E. 2000. Communities of practice and social learning systems. *Organization* 7: 225–46.

Wegner, G., Pascual, U. 2011. Cost-benefit analysis in the context of ecosystem services for human well-being: a multidisciplinary critique. *Global Environmental Change* 21: 492–504.

# 9

# INVOLVING STAKEHOLDERS

This chapter considers ways of enabling more effective knowledge exchange between policy-makers, researchers, practitioners and local communities, to anticipate, monitor and adapt to the combined effects of climate change and land degradation.

Understanding, adapting to and monitoring the interactions between climate change and land degradation requires the integration of many types of knowledge, from local to generalized; informal to formal; novice to expert; tacit and implicit to explicit; and traditional and local to scientific and universal (Raymond et al., 2010). Improved cooperation and knowledge exchange is needed between scientists and men and women in local communities, technical advisors, administrators and policy makers. Fundamentally, this is a challenge of using, and in some cases integrating, very different types of knowledge. In doing so, it may be possible to enhance the robustness of policy decisions designed to reduce the vulnerability of ecosystems and human populations to the interactions between climate change and land degradation, and develop response options that are more appropriate to the needs of local communities and can protect their livelihoods and well-being.

## 9.1 Knowledge exchange

Knowledge Exchange (KE) can be defined as generating, sharing and using knowledge through various methods appropriate to the context, purpose and participants involved (Fazey et al., 2012). KE activities range from simple transfer of information, to management of knowledge through computerized systems (Raman et al., 2011; Warner et al., 2011), through wider, complex multi-way interactions and exchange of expertise, such as in processes of adaptive co-management (Sheppard et al., 2010; Leys and Vanclay, 2011). They can also occur in formal organized, designed and intentional ways or as informal implicit processes and social learning

(Fazey *et al.*, 2012). Perceptions of what knowledge is considered to be relevant and legitimate are influenced by what knowledge is understood to be; how it is shared between those who might use it; how it is translated and/or transformed as it is shared and created; and the social context in which people produce and learn about new knowledge (Hofer, 2000; Jasanoff, 2003; Fazey *et al.*, 2012; Cook *et al.*, 2013).

Knowledge is created not just through the act of doing research, but also through social interactions and the exchange of different expertise and ways of knowing between those involved. All of these aspects are also influenced by other contextual factors (Meagher *et al.*, 2008; Morton, 2012). In individuals, learning and knowledge creation occurs through complex relationships between tacit and explicit knowledge, including through socialization (tacit to tacit), externalization (tacit to explicit), combination (explicit to explicit) and internalization (explicit to tacit) (Nonaka *et al.*, 2000). Learning and knowledge can also spread beyond individuals to groups and other social scales (Reed *et al.*, 2010). KE activities can therefore enhance knowledge sharing and knowledge use, and contribute to diverse outcomes operating at different levels, including: (i) changes in understanding (e.g. in knowledge, attitudes or ways of thinking; Kirshbaum, 2008) (conceptual impacts); (ii) changes to actual on-ground impacts (e.g. improvements in human or ecological health; Gross and Lowe, 2009; Crawford *et al.*, 2010) as a result of the implementation of new policies and practices (instrumental impacts); (iii) capacity building impacts (Meagher *et al.*, 2008); (iv) attitudinal or cultural impacts (Meagher *et al.*, 2008); and (v) process-oriented outcomes (e.g. engagement, trust, relationships, sustainability of activities; Heylings and Bravo, 2007; de Vente *et al.*, in press) (sometimes referred to as "enduring connectivity" impacts; Meagher *et al.*, 2008). These impacts may be potential, in progress (or "interim") or achieved (Meagher *et al.*, 2008).

The creation and sharing of knowledge is essentially a social process (Jasanoff, 2003) where the majority of things people learn and the beliefs they hold stem from interactions with other people, whether informally through conversation with those in their social network or via formal relationships, such as with teachers and mentors (Bandura, 1977; Sutherland *et al.*, 2004; Reed *et al.*, 2010). Even learning from written material is socially mediated, as what people learn about and trust is influenced by the society and the culture in which they are embedded (Bandura, 1977). As such, the knowledge an individual gains through engaging with research is a product of an individual's previous experience and practices, interactions, and a reflection of the cultural, social, and institutional structures of the society within which they live (Bourdieu 2001, cited in Contandriopoulos *et al.*, 2010). Consequently, the extent to which the generation of new knowledge through research becomes embodied in policy or practice is often more dependent upon the quality of the relationships of those involved than it is upon the quality of the research itself (Reed *et al.*, 2014). The social context therefore mediates the transformation of information into knowledge, and whether and how it is subsequently shared and acted upon by others (Albaek, 1995; Bourdieu 1980, 1994, cited in Contandriopoulos *et al.*, 2010).

Knowledge creation therefore needs to be understood in relation to its timeliness, access and relevance, and how this knowledge is shaped by other factors, such as group dynamics, legislation and institutions (Heylings and Bravo, 2007; Meagher *et al.*, 2008). A key and often overlooked aspect is the role of power, which influences whose voices are heard and how knowledge is created or used (Chambers, 1997; Fazey *et al.*, 2012; Williams *et al.*, 2003; Reed, 2008). While there are many conceptualizations of power in social theory, they all generally refer to the various means by which individuals and groups act and their implications for human agency (Gaventa, 1980; Valorinta *et al.*, 2011; Hatt, 2012; Avelino and Rotmans, 2009). This includes status, positional and social power, such as that mediated through pressure groups or differences in formal educational status that prevent equal participation of disadvantaged groups (Ingram and Stern, 2007). It can also include power as a "distribution of knowledge", which operates through both individual and collective action (Foucault and Gordon, 1980; Barnes, 1983). Social processes can further affect the quality of outcomes (Connick and Innes, 2003). For example, collaborative ventures with multi-stakeholder interests can produce considerable quantities of "new" knowledge but not necessarily knowledge of quality (e.g. whether it is valued or likely to be used over long time-scales). Such qualities may only emerge when there has been considerable focus on the social processes that build trust, mutual respect, legitimacy and collaborative capacity and are more widely accepted when these processes have also considered the role of power (Chambers, 1997; Connick and Innes, 2003; Kuper *et al.*, 2009).

## 9.2 Building partnerships and science-policy dialogue

Partnership building is one way to achieve long-term collaboration between different stakeholders and researchers. Partnerships are typically affected by contextual factors such as geography (e.g. determining infrastructure, subsidies and legislation), economics (e.g. size of partners), culture (e.g. finding partners with compatible values), and the existence of mutual interdependencies (e.g. technical knowledge) (Ziggers and Trienekens, 1999). Key factors in successful partnerships include: identifying clear benefits for all participants, having a good strategic fit for all partners, the involvement of management at all levels, and organizational flexibility (Hughes, 1994). For example, the UNCCD Secretariat has a "partnership framework" that seeks to build public-private partnerships on the basis of a number of clearly identified shared values and principles[1]. Developing partnerships between the UNCCD Secretariat and key stakeholders is also one of the operational objectives of the UNCCD's Comprehensive Communication Strategy[2], which aims to raise the profile and influence of soils and land in national and international policy arenas. The UNCCD's "mini guide to building partnerships"[3] compares partnership building to farming, proposing four phases: "preparing the field" (identifying stakeholders, context and barriers to developing a common vision); "sowing the seeds" (scoping objectives, identifying shared priorities, setting roles and responsibilities, assessing power dynamics and building relevant skills and expertise); "weeding and tending the growing plants" (needs assessment, strategy development, co-ordination, monitoring and evaluation); and "harvesting" (institutionalizing partnerships, maintaining linkages, reporting progress and achieving continuity). Dyer *et al.* (2014) build on this drawing on lessons from multi-stakeholder climate compatible development projects in sub-Saharan Africa, to propose that good practice in partnership building and multi-stakeholder engagement should ensure projects are appropriate and relevant to local needs, by engaging communities from the outset.

There are increasing calls for environmental policy decisions to be based on research evidence and monitoring at a variety of scales (e.g. UNCCD, 1994; Sutherland *et al.*, 2004; Pullin *et al.*, 2004; MA, 2005; Reid *et al.*, 2006; WOCAT, 2007; Jessop *et al.*, 2008) but this remains a challenge. Studies of science-policy communication have tended to focus on the perceptions of researchers e.g. surveys of authors publishing in conservation journals about whether they perceive their recommendations have been implemented (Flaspohler *et al.*, 2000; Ormerod *et al.*, 2002). Few studies have focused on the perceptions of those responsible for using this evidence in decision making, or have analysed the pathways through which research evidence and monitoring data reaches policy makers, or if is transformed or blocked by the actors involved (Knight *et al.*, 2008; Spierenburg, 2012). Evidence can often be distorted as it is passed from person to person through social networks, and is sometimes misappropriated to achieve the goals of special interest groups. Robust evidence may be overlooked and more flimsy findings may gain traction with decision-makers who do not always have the time or expertise to critically interrogate its theoretical, methodological or empirical basis. Instead it is often easier to judge evidence on basis of the trustworthiness of its source, whether that be the quality of the journal it is published in or the credentials of the person who communicates it.

The situation is complicated further by the highly fractured nature of the current knowledge base across different institutions and academic disciplines, combined with structural and procedural barriers that prevent the flow of knowledge between those who are monitoring land degradation and climate change at different scales (Reed *et al.*, 2013). With limited coordination of monitoring activities or integration of data across scales, those working at national and international levels are rarely able to tap into the data and expertise held by those who manage the land. In turn, land managers rarely see the benefits of national and international monitoring programmes (Reed *et al.*, 2006). However, if knowledge about monitoring and assessment can be managed more effectively, it may be possible to provide a more robust evidence-base that can support more sustainable land management policies and practices to tackle land degradation and climate change (Raymond *et al.*, 2010). Scientific literature provides valuable evidence, and a means of generating certain types of data from monitoring and evaluation, but if we are to capture the dynamic, context-dependent and value-laden nature of land degradation, we cannot overlook the equally valuable but often unrecognized knowledge of local people and the contributions of the NGOs, CBOs and CSOs that work with them. To achieve KE across scales, it will be necessary to better understand the structure of the social and organizational networks that exist around those generating and using knowledge about land degradation and climate change. In this way, it may be possible to identify knowledge brokers and boundary organizations that can efficiently exchange knowledge between communities of actors who would otherwise have little communication with one another. Partly, this is being addressed by increasing efforts to share knowledge between those involved in the implementation of the Rio conventions (see Chapter 2 and the example of the JLG), but key challenges remain in facilitating KE between actors operating at international and local scales. More research is needed to understand the barriers to communication and KE, and how these might be overcome in future.

## 9.3 Who needs to exchange knowledge about climate change and land degradation?

Given the complexity and knowledge gaps around the links between climate change and land degradation that have been highlighted in this report, it is essential to pool knowledge from different sources to better understand the processes involved, the likely response options and to be able to effectively monitor responses. Partly KE needs to be facilitated through the development of cross-institutional initiatives and mechanisms for evidence-based policy, including Science-Policy Interfaces like the IPCC, IPBES and the newly established SPI and Scientific Knowledge Brokering Portal for the UNCCD, as well as through assessments like the MA and the IPBES LDRA. (See also Akhtar-Schuster et al. (2011) and Chasek et al. (2011) for more detailed discussions of opportunities for horizontal knowledge management at national and international levels).

Partly, KE needs to be facilitated between local communities, civil society and policy makers at national and international scales. The UNCCD is anchored in the recognition that the top-down, science-led technology transfer paradigm is inadequate for combating desertification. It was argued that by tapping into local and traditional knowledge, more complete information could lead to more robust solutions to environmental problems. Compared to the other Rio Conventions, the UNCCD affords the most attention to the participation of civil society and local communities in its text. However, there is no formal mechanism to ensure local and traditional knowledge is taken into account in UNCCD processes and negotiations. Locally held knowledge has typically entered the international UNCCD process via: (i) participation of local communities representatives in the process via CSOs and NGOs, farmers' associations, unions and local authorities (though each have relatively little power in the policy arena and do not necessarily represent the diversity of local knowledge in their constituencies); and (ii) national UNCCD processes such as NAP consultations and interactions with science and technology correspondents, UNCCD national reporting processes and National Coordinating Bodies (where knowledge from the local level can be included in National Reports to be submitted to the CRICs and COPs, or can be taken along in negotiations by the delegations of the various Parties). However, there is no stock-taking mechanism for national reports to gather relevant experiences and knowledge and present it in a practical way for upscaling and/or dissemination. Current pathways depend largely on the willingness of individuals and Parties to take this knowledge into account (DESIRE/DryNet/eniD, 2008).

Knowledge exchange also needs to be facilitated between researchers and stakeholders affected by climate change and land degradation. By combining scientific knowledge with locally held knowledge through participatory research, it may be possible to enable enhanced adoption of innovations that can enable adaptation in different social and environmental contexts. Local populations are often well placed to collect data and take part in monitoring and evaluation.

## 9.4 Engaging with stakeholders to tackle land degradation and climate change

Climate change and land degradation are highly complex processes, affecting many different stakeholders at different scales. Traditional top-down approaches to environmental challenges such as climate change and land degradation often face serious problems when it comes to implementation (Cramb *et al.*, 1999; Knill and Lenschow, 2000). Often, these problems can be attributed to the lack of ownership over the process amongst those who have the power to implement decisions (e.g. state actors or affected citizens and land owners), leading to low rates of acceptance. This may then lead to these groups delaying or preventing the implementation of decisions, in order to preserve their own interests.

Attempts to tackle these processes, and the interactions between them, therefore require engagement with diverse, and often conflicting stakeholder priorities, especially where response options lead to trade-offs between different ecosystem services. Often, this is a trade-off between short-term provisioning services (e.g. crop and animal production or extractive uses of forests) upon which the resource-dependent poor often depend for their livelihoods, versus the protection and enhancement of regulating and supporting services (such as nutrient cycling and soil formation), which have the potential to reverse land degradation, contribute to LDN and enhance resilience to climate change.

Participatory approaches are often sought to address these conflicts. Indeed, it has been claimed that more bottom-up, participatory approaches to tackling climate change and land degradation have the capacity to reduce conflict, build trust and facilitate learning amongst stakeholders, who are then more likely to support adaptation in the long term (e.g. Beierle, 2002; Reed, 2008). However, participatory approaches by definition seek and value a plurality of views, rather than seeking a single evidence-based solution. Taking a participatory approach, scientific evidence therefore becomes one of many lines of arguments in a multi-stakeholder discourse about different options for responding to the effects of climate change and land degradation. The very nature of participation means that conflicting interests will be made explicit and brought to the fore, and local knowledge will be evaluated alongside scientific knowledge.

For this to be effective, participatory processes need to facilitate deliberation between participants over the options to tackle land degradation and climate change.

Deliberation occurs when people search for information and gain knowledge to form reasoned opinions that they can express in reasoned dialogue with others, to identify or evaluate options and then apply insights from the deliberation to determine well informed contextual values and preferences in relation to these options (Kenter *et al.*, 2014). Deliberation also enables different groups of people within society to learn from one another through their interactions with each other (this is sometimes termed "social learning"; see Reed *et al.*, 2010). As such, the value of deliberation is often not (or not just) in sharing values and reaching consensus, but also in appreciating the reasons behind other people's values, helping people to be able to "live with" decisions that emerge from the process, whether they agree with the outcome or not. A variety of participatory techniques can facilitate this (Box 9.1).

---

## BOX 9.1: PARTICIPATORY TECHNIQUES

Broadly speaking, there are two types of participatory methods: "deliberative" techniques enable participants to exchange and consider evidence together and negotiate; while "analytical-deliberative" techniques are more structured, integrating deliberation with analytical tools such as in Deliberative Monetary Valuation (DMV).

Deliberation may be used at various points in decision-making processes, for example:

- *exploratory phase*: understanding the sorts of challenges stakeholders are facing that the decision might be able to address; scoping the objectives and approach to ensure the outcomes of the decision are as relevant as possible to everyone involved in the decision;
- *evidence collection and analysis*: it may be useful to gather evidence with stakeholders through deliberation to elicit shared values, appraise options and better understand attitudes, perceptions and likely reactions to potential decisions among different groups; and
- *interpretation of evidence*: whether evidence comes from stakeholders or other sources, it may be useful to engage stakeholders in the interpretation of evidence, making links and contributions to issues that might otherwise have been overlooked.

There are a number of mechanisms that can be used to facilitate deliberation. Broadly these can be categorized as methods for:

- *Opening up* dialogue and gathering information with stakeholders about issues linked to the decision, for example:
  - *brainstorming* (getting participants to think rapidly and express ideas in short phrases to come up with new and creative ideas);

- ○ *metaplanning* (participants are given a number of post-it notes, asked to write one idea per post-it and place it on the wall, grouped next to ideas that sound similar, so ideas cluster); and
- ○ listing techniques such as a *carousel* where questions are arranged in stations around the room and groups move around the stations, each group with a different coloured pen, at timed intervals, until they arrive back at their starting station and can read what other groups added to their initial ideas.

- • *Exploring* issues in greater depth with stakeholders, for example:
  - ○ *mind-mapping®* techniques (also known as concept mapping, spray, spaghetti and spider diagrams) can be a useful way to quickly capture and link ideas with stakeholders;
  - ○ *SWOT analysis* encourages people to think systematically about the strengths, weaknesses, opportunities and threats as they pertain to the decision under consideration; and
  - ○ for issues that have a strong temporal dimension or for project planning with stakeholders, *timelines* can be used to help structure discussion in relation to historical or planned/hoped for future events.

- • *Closing down* options and deciding on actions, for example:
  - ○ *ranking* can be used to place ideas in rank order. Getting consensus amongst participants for a particular ranking can be challenging, but the discussions it stimulates may be revealing;
  - ○ *prioritization* differs from ranking by enabling participants to express the strength of their feeling towards a particular option rather than simply ranking an idea as better or worse than another idea e.g. by assigning sticky dots or stones (if working outside) to different ideas or options;
  - ○ establishing a *verdict* is a qualitative way of appraising or choosing a preferred option;
  - ○ *willingness to pay* (WTP) to achieve a certain outcome or for putting a policy or a form of management into place (e.g. establishing a protected area) can be deliberated at the individual scale (what an individual is WTP), or as a "fair price" (identifying what is a fair price to pay for a member of the public or a certain group or community) or as social WTP (how much society/government should pay). The advantage of using amounts of money is that the relative importance of different things can be easily and pragmatically compared; and
  - ○ *multi-criteria (decision) analysis* combines ratings across different dimensions of value (the criteria). It allows economic, social and environmental criteria, including competing priorities, to be systematically evaluated by groups of people.

According to Kenter *et al.* (2014), the extent to which deliberation or social learning leads to greater sharing of values especially depends on:

1. the diversity of initial values in the group (if everyone in the group has quite similar values, it may be easier to discover or reach shared values, but there is less scope for changes in values to occur as a result of deliberation);
2. how effectively values are made explicit in the deliberation (this may be easier for some participants to do than others);
3. how effectively the process is designed and facilitated (in particular to make values explicit and to manage power dynamics); and
4. the length of time over which deliberation occurs.

## 9.5 What makes participation work?

If stakeholder participation and deliberation is conducted effectively, it may be possible to evaluate and consider different sources of knowledge equally, and facilitate decisions that are supported by the majority of stakeholders. However, there are many examples of participatory approaches to climate change and land degradation that have failed to deliver the intended beneficial environmental or social outcomes. It is often unclear why different participatory processes in different contexts lead to such different outcomes. Despite many local case studies of participatory approaches to climate change and land degradation adaptation, there have been few attempts to generalize from these experiences to explain how and why participatory approaches sometimes work, and sometimes are ineffective and exacerbate local conflicts (de Vente *et al.*, in press). We therefore propose a theory of participation that can explain why participation in initiatives to tackle land degradation and climate change sometimes work and sometimes do not. By elucidating the theoretical mechanisms through which participation operates, we hope to provide a theoretical foundation for good practice participation.

The theoretical framework is composed of four guiding principles that seek to explain why stakeholder participation in decisions about environmental management to mitigate land degradation and climate change may or may not lead to desired outcomes. If participation is defined as the active involvement of all Parties who are affected by or can affect a decision, then the extent to which participation contributes constructively to the decision-making process will depend on:

1. the context in which the participation occurs;
2. the design of the participatory process;
3. the extent to which the facilitator/mediator is able to manage power dynamics within this design and enable deliberation between participants; and
4. the length of time and spatial scales over which the participatory process occurs.

The first principle relates to the extent to which participation is significantly affected by the socio-economic, cultural and institutional contexts within which it is enacted.

There are a number of contexts in which participation may not be appropriate in tackling land degradation and climate change, or in which potential participants have little previous experience, for example: where there is widespread apathy and disengagement among stakeholders, making it difficult to mobilize participation; where there is an autocratic culture (i.e. with little decision-making autonomy for individuals), as in former communist states (Stringer *et al.*, 2009); where there are significant power imbalances between participants; or where some or all participants do not really have decision-making power (Reed, 2008; de Vente *et al.*, in press).

The second principle is that there are a small number of process-design variables that will increase the likelihood that participation contributes towards decision-making processes across a range of contexts. De Vente *et al.* (in press) present evidence based on an analysis of different types of participatory process in similar contexts versus similar process designs in very different contexts. Their study suggests that participatory processes that deliver beneficial environmental and social outcomes are more likely to include the legitimate representation of stakeholders and the provision of information and decision-making power to all participants. Although participatory processes initiated or facilitated by government bodies led to significantly less trust, information gain, learning and less flexible solutions, decisions in these processes were more likely to be accepted and implemented by participants without being contested.

This then leads to the third principle: that a well-designed process needs to be effectively facilitated to balance power dynamics between participants (de Vente *et al.*, in press). The fourth principle notes that the success of participation is highly scale-dependent over space and time. The extent to which participation facilitates learning, shapes the values of participants and tends towards consensus rather than conflict, is highly dependent on the temporal scales over which participation occurs. Assuming that values are made explicit as part of the deliberative process, deliberation within participatory processes may alter contextual values over short timescales (e.g. a single workshop), but more deeply held values and beliefs are likely to require engagement over much longer periods of time, potentially to generational timescales, for more deeply engrained conflicts that have become embedded in the cultural norms of a society to be altered.

The extent to which participation can deliver beneficial environmental and social outcomes in the context of addressing the interlinked challenges of land degradation and climate change may also depend upon matching the participation of relevant stakeholders to the spatial scales at which a decision is being made or over which a conflict is operating. For national and international environmental decision-making processes and conflicts, this can be challenging. Broadly, there are two mechanisms through which participation can deliver outcomes at different spatial scales: replication and social learning. In reality, a mix of these two processes probably occurs. Following the replication approach, participatory processes are replicated across a particular spatial unit (e.g. habitat, region or nation).

Following the social learning approach, participation is designed to enable knowledge and attitudes arising from a participatory process to diffuse through the social networks of those directly involved, and these representatives draw on their shared values and the knowledge of those they are connected to in their social network. Again, for this to be effective, each of the principles will need to have been fulfilled, in particular ensuring effective representation of different stakeholder interests. More sophisticated approaches using Social Network Analysis may be used to identify individuals or organizations who are central in social networks in terms of their connectivity and trust. However, the notion of homophily suggests that with the exception of certain key individuals who act as knowledge brokers and connect disparate social networks, most stakeholders are likely to diffuse and represent the values and knowledge of others who are similar to them. This creates a challenging context for participation within the context of DLDD and climate change, as although many of the stakeholders are the same for the two issues, there are also a number of different groups who have a stake in only one of these issues.

The theoretical principles outlined above will now be used to propose good practice in participation that is both theoretically robust, and based on empirical experience of what works in practice.

## 9.6 Good practice participation

### 9.6.1 *The right people*

A diverse group of well-informed people from different backgrounds is likely to provide the most relevant and innovative ideas to tackle the combined challenges of climate change and land degradation. There are two tasks that need to be performed: identifying the right people for inclusion in the participatory process; and actually getting these people engaged.

First, it is necessary to identify the most relevant individuals and organizations that can represent the full range of interests in the decision-making process. For issues like climate change and land degradation, these stakeholders may operate at local, national and international levels, and it is therefore important to define the scale at which participation is sought. Where multiple scales are sought it may be necessary to facilitate different processes at these different scales, given the likely differences in location (and hence travel distances) between stakeholders, and their different interests in the issues. If the focus is on only one scale, it is still often important to be aware of and link to stakeholder interests at different scales, to ensure options emerging from the process are viable. If key actors are missing from the participatory process, then they are likely to question the legitimacy of the process and potentially block progress towards implementing outcomes. If possible, once representation has been achieved, it can be useful to consider whether the right people have been invited from the organizations that are represented. If these people do not have decision-making power and have to refer any decision back to their superiors, it will be difficult to reach any agreement as part of the participatory process. Where possible, trying to target a few representatives who are known for being innovative and creative can help the participatory process achieve more creative solutions to climate change and land degradation.

Second, it is necessary to actually bring these people into the participatory process. A participatory process that has identified all the key players but fails to engage them is likely to lead to biased outcomes with low acceptance and implementation. Partly this is about effective communication, and making involvement attractive and easy for all participants. For some participants, this may be about practical considerations (e.g. avoiding meetings at certain times of day or certain seasons) or financial or other types of incentives (e.g. payments for participation, offering meals and opportunities to network). In the international context there are many bureaucratic hurdles to contend with as well, such as the need to obtain visas for some participants to attend meetings in foreign countries. For many participants, they simply need to believe there is a high probability that engaging in the process will lead to direct benefits (e.g. access to land, compensation, etc.). By working through existing trusted contacts and networks, it may be possible to reach and convince potential participants who may not otherwise have been contacted and involved.

### 9.6.2 The right atmosphere

Creating an open and respectful environment needs to start at the very beginning and continue throughout people's engagement with the process. There is little point in having the right people engaged in the process if some dominate and others feel powerless to speak. Working with a professional, independent mediator can help create the right atmosphere, enabling everyone to have an equal say. They can manage conflicts as they arise, and they often have tools that can get lots of information from people very quickly, for them to think about critically together.

The methods that a mediator employs can go a long way towards creating an atmosphere of trust. Methods must be adapted to the socio-cultural context of the participatory process e.g. avoiding methods that require participants to read or write in groups that might include illiterate participants. Depending on the power dynamics of the group, methods may need to be employed that rebalance power between participants to avoid marginalizing the voices of the less powerful. There is evidence that actors who are marginalized during decision-making can delay or prevent implementation through litigation.

### 9.6.3 Making it relevant

Finally, it is important to make the participatory process as relevant as possible for all participants. Partly this is about the content and focus of the participatory process, and how this focus is derived. Partly it is about the perceived credibility of the process, and the likelihood that it will lead to beneficial change. If participants do not perceive that the process has credibility to affect the issues that concern them, then they will view it as irrelevant and not engage with it. Negotiating a set of ambitious but achievable goals with all participants from the outset can help demonstrate that their participation is likely to make a real difference. If the goals are developed through dialogue between participants (making trade-offs where necessary), they are more likely to take ownership of the process, partnership building will be more likely, and the outcomes are more likely to be more relevant to stakeholder needs and priorities, motivating their ongoing active engagement.

Once the content and focus of participation has been negotiated and agreed by all Parties, the approach to participation needs to be made as relevant as possible to all participants. For example, language can sometimes be used to erect barriers between different groups who each have their own exclusive vocabulary, so it is important to try and use language that is familiar and impartial to all Parties. When dealing with complex, intangible concepts (such as biodiversity), it may be necessary to focus on aspects that are more tangible to participants (such as indicator species that have known uses for humans).

## 9.7 Synthesis

Understanding, adapting to and monitoring the interactions between climate change and land degradation requires engagement with the widest possible range of affected stakeholders, and the integration of many different types of knowledge. Given the number of knowledge gaps around the links between climate change and land degradation that we have highlighted, there is an important role for the generation of new knowledge, as well as the sharing and mobilization of existing and new knowledge from different sources. Knowledge exchange needs to be facilitated through the development of cross-institutional initiatives and mechanisms for evidence-based policy, such as the IPCC, IPBES, and the UNCCD's SPI, and assessments like the MA and the LDRA. By combining different sources of knowledge and engaging with stakeholders in this way, it may be possible to enhance adoption of innovations that can enable adaptation in different social and environmental contexts.

Any attempt to integrate knowledge from such different sources, and to negotiate between divergent perspectives and priorities, is likely to lead to conflicts of interest. This is particularly likely where proposed adaptation options are anticipated to lead to trade-offs between ecosystem services. For example, there is often a trade-off between short-term provisioning services (such as crop and animal production or extractive uses of forests) versus the protection and enhancement of regulating and supporting services (such as nutrient cycling and soil formation), which have the potential to reverse land degradation and enhance resilience to climate change.

Participatory approaches are often sought to address these conflicts. This chapter has presented evidence that good practice participation has the potential to reduce conflict, build trust and facilitate learning amongst stakeholders, who may then be more likely to support project goals and implement decisions in the long-term. Key principles of good practice relate to the appropriate representation of stakeholder interests, the management of power dynamics and the relevance of the process to stakeholder needs and priorities.

## Notes

1 www.unccd.int/en/about-the-convention/the-bodies/the-cop/cop11/Pages/Call-for-interests-in-the-UNCCD-secretariat-and-the-private-sector-partnership.aspx
2 www.unccd.int/Lists/SiteDocumentLibrary/convention/css.pdf
3 www.unccd.int/Lists/SiteDocumentLibrary/Partnerships/Mini_Guide.pdf

## References

Akhtar-Schuster, M., Thomas, R.J., Stringer, L.C., Chasek, P., Seely, M.K. 2011. Improving the enabling environment to combat land degradation: institutional, financial, legal and science-policy challenges and solutions. *Land Degradation and Development* 22: 299–312.

Albaek, E. 1995. Between knowledge and power: utilization of social science in public policy making. *Policy Sciences* 28: 79–100.

Avelino, F., Rotmans, J. 2009. Power in transition: an interdisciplinary framework to study power in relation to structural change. *European Journal of Social Theory* 12: 543–569.

Bandura, A. 1977. *Social Learning Theory*. General Learning Press: New York.

Barnes, B. 1983. Social-life as bootstrapped induction. *Sociology* 17: 524–545.

Beierle, T.C. 2002. The quality of stakeholder-based decisions. *Risk Analysis* 22: 739–749.

Bourdieu, P. 1980. *Le sens pratique*. Paris: Éditions de Minuit.

Bourdieu, P. 1994. Un acte désintéressé est-il possible? In: *Raisons prtiques sur la théorie de l'action*, Bourdieu, P. (ed.). 149–67. Paris: Éditions du Seuil.

Bourdieu, P. 2001. Le mystère du ministère: Des volontés particulières à la "volonté générale". *Actes de la recherche en science sociale* 140:7–11.

Chambers, R. 1997. *Whose Reality Counts? Putting the first last*. ITDG Publishing: London.

Chasek, P., Essahli, W., Akhtar-Schuster, M., Stringer, L.C., Thomas, R. 2011. Integrated land degradation monitoring and assessment: horizontal knowledge management at the national and international levels. *Land Degradation & Development* 22: 272–284.

Connick, S., Innes, J.E. 2003. Outcomes of collaborative water policy making: applying complexity thinking to evaluation. *Journal of Environmental Planning and Management* 46: 177–197.

Contandriopoulos, D., Lemire, M., Denis, J.-L., Tremblay, É. 2010. Knowledge exchange processes in organizations and policy arenas: a narrative systematic review of the literature. *The Milbank Quarterly* 88: 444–83.

Cook, B.R., Atkinson, M., Chalmers, H., Comins, L., Cooksley, S., Deans, N., Fazey, I., Fenemor, A., Kesby, M., Litke, S., Marshall, D., Spray, C. 2013. Interrogating participatory catchment organisations: cases from Canada, New Zealand, Scotland and the Scottish-English Borderlands. *Geographical Journal* 179: 234–247.

Cramb, R.A., Garcia, J.N.M., Gerrits, R.V., Saguiguit, G.C. 1999. Smallholder adoption of soil conservation technologies: evidence from upland projects in the Philippines. *Land Degradation & Development* 10: 405–423.

Crawford, B., *et al.* 2010. Small scale fisheries management: lessons from cockle harvesters in Nicaragua and Tanzania. *Coastal Management* 38: 195–215.

DESIRE/DryNet/eniD. 2008. Is the UNCCD stuck in a knowledge traffic jam? Discussion paper for CRIC 7 by DESIRE/Drynet/eniD. Available online at: www.desire-his.eu/index.php/en/global-level/220-is-the-unccd-stuck-in-a-knowledge-traffic-jam.

De Vente, J., Reed, M.S., Stringer, L.C., Valente, S., Newig, J. (in press). How does the context and design of participatory decision-making processes affect their outcomes? Evidence from sustainable land management in global drylands. *Ecology & Society*.

Dyer, J., Stringer, L.C., Dougill, A.J., Leventon, J., Nshimbi, M., Chama, F., Kafwifwi, A., Muledi, J.I., Kaumbu, J.M., Falcao, M., Muhorro, S., Munyemba, F., Kalaba, G.M., Syampungani, S. 2014. Assessing participatory practices in community-based natural resource management: experiences in community engagement from southern Africa. *Journal of Environmental Management* 137: 137–145.

Fazey, I., Evely, A.C., Reed, M.S., Stringer, L.C., Kruijsen, J., White, P.C.L., Newsham, A., Jin, L., Cortazzi, M., Phillopson, J., Blackstock, K., Entwhistle, N., Sheate, W., Armstrong, F., Blackmore, C., Fazey, J., Ingram, J., Gregson, J., Lowe, P., Morton, S., Trevitt, C. 2012. Knowledge exchange: a research agenda for environmental management. *Environmental Conservation* 40: 19–36

Flaspohler, D.J., Bub, B.R., Kaplin, B.A. 2000. Application of conservation biology research to management. *Conservation Biology* 14: 1898–1902.

Foucault, M., Gordon, C. 1980. *Power/knowledge: Selected interviews and other writings, 1972–1977*. Harvester Press: New York.

Gaventa, J. 1980. *Power and Powerlessness: Quiescence and rebellion in an Appalachian valley*. Clarendon Press: Oxford.

Gross, D.P., Lowe, A. 2009. Evaluation of a knowledge translation initiative for physical therapists treating patients with work disability. *Disability and Rehabilitation* 31: 871–879.

Hatt, K. 2012. Social attractors: a proposal to enhance 'resilience thinking' about the social. *Society and Natural Resources* 26: 30–43.

Heylings, P., Bravo, M. 2007. Evaluating governance: a process for understanding how co-management is functioning, and why, in the Galapagos Marine Reserve. *Ocean and Coastal Management* 50: 174–208.

Hofer, B.K. 2000. Dimensionality and disciplinary differences in personal epistemology. *Contemporary Educational Psychology* 25: 378–405.

Hughes, D. 1994. *Breaking with Tradition: Building partnerships and alliances in the European food industry*. Wye College Press: Wye.

Ingram, H. Stern, P. 2007. *Research and Networks for Decision Support in the NOAA Sectoral Applications Research Program*. National Academies Press: Washington, D.C., USA.

Jasanoff, S. 2003. Technologies of humility: citizen participation in governing science. *Minerva* 41: 223–244.

Jessop B., Brenner N., Jones M. 2008. Theorizing socio-spatial relations. *Environment and Planning D: Society and Space* 26: 389–401.

Kenter, J.O., Reed, M.S., Irvine, K.N., O'Brien, E., Brady, E., Bryce, R., Christie, M., Church, A., Cooper, N., Davies, A., Hockley, N., Fazey, I., Jobstvogt, N., Molloy, C., Orchard-Webb, J., Ravenscroft, N., Ryan, M., Watson, V. 2014. *UK National Ecosystem Assessment Follow-on Phase, Technical Report: Shared, plural and cultural values of ecosystems*. UNEP-WCMC: Cambridge.

Kirshbaum, M. 2008. Translation to practice: a randomised, controlled study of an evidence-based booklet for breast-care nurses in the United Kingdom. *Worldviews on Evidence-Based Nursing* 5: 60–74.

Knight, A.T., Cowling, R.M., Rouget, M., Balmford, A., Lombard, A.T., Campbell, B.M. 2008. Knowing but not doing: selecting priority conservation areas and the research-implementation gap. *Conservation Biology* 22/3: 610–617.

Knill, C., Lenschow, A. (eds). 2000. *Implementing EU Environmental Policy: New directions and old problems*. Manchester University Press: London.

Kuper, M., Dionnet, M., Hammani, A., Bekkar, Y., Garin, P., Bluemling, B. 2009. Supporting the shift from state water to community water: lessons from a social learning approach to designing joint irrigation projects in Morocco. *Ecology and Society* 14(1): 19.

Leys, A.J., Vanclay, J.K. 2011. Social learning: a knowledge and capacity building approach for adaptive co-management of contested landscapes. *Land Use Policy* 28: 574–584.

MA (Millennium Ecosystem Assessment) 2005. *Ecosystems and Human Well-being: Policy responses*. Island Press: Washington, D.C., USA.

Meagher, L., Lyall, C., Nutley, S. 2008. Flows of knowledge, expertise and influence: a method for assessing policy and practice impacts from social science research. *Research Evaluation* 17: 163–173.

Morton, S. 2012. *Exploring and Assessing Social Research Impact: A case study of a research partnership's impacts on policy and practice*. University of Edinburgh: Edinburgh, UK.

Nonaka, I., Toyama, R., Konno, N. 2000. SECI, Ba and leadership: a unified model of dynamic knowledge creation. *Long Range Planning* 33: 5–34.

Ormerod, S.J., Barlow, N.D., Marshall, E.J.P., Kerby, G. 2002. The uptake of applied ecology. *Journal of Applied Ecology* 39: 1–7.

Pullin, A.S., Knight, T.M., Stone, D.A., Charman, K. 2004. Do conservation managers use scientific evidence to support their decision-making? *Biological Conservation* 119: 245–252.

Raman, M., Dorasamy, M., Muthaiyah, S., Kaliannan, M., Muthuveloo, R. 2011. Knowledge management for social workers involved in disaster planning and response in Malaysia: an action research approach. *Systemic Practice and Action Research* 24: 261–272.

Raymond, C.M., Fazey, I., Reed, M.S., Stringer, L.C., Robinson, G.M., Evely, A.C. 2010. Integrating local and scientific knowledge for environmental management: from products to processes. *Journal of Environmental Management* 91: 1766–1777.

Reed, M.S., Fraser, E.D.G., Dougill, A.J. 2006. An adaptive learning process for developing and applying sustainability indicators with local communities. *Ecological Economics* 59: 406–418.

Reed, M.S. 2008. Stakeholder participation for environmental management: a literature review. *Biological Conservation* 141: 2417–2431.

Reed, M.S., Evely, A.C., Cundill, G., Fazey, I., Glass, J., Laing, A., Newig, J., Parrish, B., Prell, C., Raymond, C., Stringer, L.C. 2010. What is social learning? *Ecology & Society* 15: r1. Available online at: www.ecologyandsociety.org/vol15/iss4/resp1/.

Reed, M.S., Fazey, I., Stringer, L.C., Raymond, C.M., Akhtar-Schuster, M., Begni, G., Bigas, H., Brehm, S., Briggs, J., Bryce, R., Buckmaster, S., Chanda, R., Davies, J., Diez, E., Essahli, W., Evely, A., Geeson, N., Hartmann, I., Holden, J., Hubacek, K., Ioris, I., Kruger, B., Laureano, P., Phillipson, J., Prell, C., Quinn, C.H., Reeves, A.D., Seely, M., Thomas, R., van der Werff Ten Bosch, M.J., Vergunst, P., Wagner, L. 2013. Knowledge management for land degradation monitoring and assessment: an analysis of contemporary thinking. *Land Degradation & Development* 24: 307–322.

Reed, M.S., Stringer, L.C., Fazey, I., Evely, A.C., Kruijsen, J. 2014. Five principles for the practice of knowledge exchange in environmental management. *Journal of Environmental Management* 146: 337–345.

Sheppard, D.J., Moehrenschlager, A., McPherson, J.M., Mason, J.J. 2010. Ten years of adaptive community-governed conservation: evaluating biodiversity protection and poverty alleviation in a West African hippopotamus reserve. *Environmental Conservation* 37: 270–282.

Spierenburg, M. 2012. Getting the message across: biodiversity science and policy interfaces – a review. *GAIA* 21/2: 125–134.

Stringer, L.C., Scrieciu, S.S., Reed, M.S. 2009. Biodiversity, land degradation, and climate change: participatory planning in Romania. *Applied Geography* 29: 77–90.

Sutherland, W.J., Pullin, A.S., Dolman, P.M., Knight, T.M. 2004. The need for evidence-based conservation. *Trends in Ecology & Evolution* 19: 305–308.

UNCCD. 1994. *United Nations Convention to Combat Desertification in Those Countries Experiencing Serious Drought and/or Desertification Particularly in Africa: Text with annexes.* UNEP: Nairobi.

Valorinta, M., Schildt, H., Lamberg, J.A. 2011. Path dependence of power relations, path-breaking change and technological adaptation. *Industry and Innovation* 18: 765–790.

Warner, G., *et al.* 2011. Advancing coordinated care in four provincial healthcare systems: evaluating a knowledge-exchange intervention. *Healthcare Policy* 7: 80–94.

Williams, G., Veron, R., Corbridge, S., Srivastava, M. 2003. Participation and power: poor people's engagement with India's employment assurance scheme. *Development and Change* 34: 163–192.

WOCAT. 2007. *Where the Land is Greener: Case studies and analysis of soil and water conservation initiatives worldwide.* Liniger, H., Critchley, W. (eds). CTA: Wageningen.

Ziggers, G.W., Trienekens, J. 1999. Quality assurance in food and agribusiness supply chains: developing successful partnerships. *International Journal of Production Economics* 60–61: 271–279.

# 10
# CONCLUSION

Although it is well recognized that climate change and land degradation present major challenges to livelihoods and human well-being, little attention has been paid to the way climate change may combine with land degradation in the future to create new and potentially unexpected challenges. The likely impacts of climate change and land degradation have typically been examined separately, and in isolation from their socio-economic and governance contexts. Although this approach has been widely critiqued (e.g. Blaikie *et al.*, 1994; Bohle, 2001; Hilhorst and Bankoff, 2004; Reed *et al.*, 2011), there have been few attempts at more integrated assessments. This book is one of the first attempts to consider how the land management and climate change communities can work together to better anticipate, asses, and adapt to the combined effects of climate change and land degradation.

We have taken an interdisciplinary and integrated systems approach to climate change and land degradation as interlinked concepts that have both biophysical and human drivers, impacts and responses. As such, the likely effects and most appropriate response options can only be determined by considering both biophysical and socio-economic data, interpreted in relation to qualitative (and often subjective) information about the livelihood strategies of those affected. This is because land degradation can only be defined in relation to the objectives of those using the land (Warren, 2002), so one form of environmental change might represent degradation to one land user, whilst it may represent a livelihood opportunity to another. This may present an even greater social science challenge if definitions of land degradation are extended to cover any medium- or long-term, permanent decline in the provision of ecosystem services more generally, including cultural services (Reed *et al.*, 2015).

In areas affected by DLDD, it is particularly challenging to anticipate how climate change and land degradation will interact, because (unlike mesic systems) drylands experience pulses in the availability of many resources, especially water.

Drylands consequently often rely upon resource subsidies from surrounding areas. These features occur for both natural and human-managed dryland ecosystems. The pulsed nature of natural drylands is well documented (e.g. Schwinning et al., 2004 and the rest of special issue *Oecologia* 141(2)), and management activities are often directed towards mitigating pulsed resource availability (e.g. livestock wells/boreholes) or subsidizing resources (e.g. irrigation systems). The pulsed nature of drylands means that climate change and DLDD's interactive effects will be a function of deviations from current climate/management regimes, and the water-limited nature of drylands means that climate change and DLDD's interactive effects also depend on the ability of surrounding systems to be sustainably managed, such that they can continue to export resources.

## 10.1 Key vulnerabilities to the combined effects of climate change and land degradation

Using our conceptual and methodological frameworks (Figures 3.3 and 3.4), it is possible to identify a number of key vulnerabilities to the combined effects of climate change and land degradation at a more generalized, global level. First, it is important to recognize that exposure to climate change varies globally, with different regional projections of changes in temperature, rainfall and sea-level rise. Likewise, different regions are exposed to different types and levels of land degradation, and it is impossible to assess the vulnerability of populations and ecosystems to either climate change or land degradation solely on the basis of these differing levels of exposure. However, assessments of current and likely future exposure to climate change and land degradation can provide an important basis for assessing the sensitivity of social-ecological systems to those changes.

In particular, areas already experiencing land degradation are likely to be exposed to potentially damaging interactions with climate change, where extreme weather events such as increased droughts or heavy rainfall events exacerbate wind or water erosion and (with unchanged agricultural practices) contribute towards further reductions in biomass or physical and chemical degradation of soils. Further research is needed to combine assessments of land degradation and climate change impacts, to better understand the extent to which different ecosystems and populations around the world are likely to be exposed to important combinations of changes resulting from both processes. The full extent to which this exposure to risks from climate change and land degradation leads to negative impacts on ecosystems and populations can only be understood by considering the relative sensitivity of different systems to the interactions between climate change and land degradation.

Assessing the sensitivity of ecosystems and populations to the combined effects of climate change and land degradation is in part a biophysical challenge. It is necessary to understand how land degradation processes, such as water and wind erosion and physical and chemical degradation of soils, might interact with changes in soil temperature, precipitation (amount, intensity and patterns), humidity, atmospheric $CO_2$ concentrations and evapotranspiration rates. Given the high temperatures and limited rainfall already experienced in drylands, these regions are likely to be particularly sensitive to the effects of climate-induced changes in temperature and moisture, particularly when combined with degradation-induced reductions in soil organic matter, biomass and soil fertility. These processes may in some cases be self-reinforcing, leading to feedbacks between climate change and land degradation, for example when land degradation via the loss of terrestrial carbon stores from soils and vegetation leads to climate warming, or when the albedo effect of degradation-induced reductions in vegetation cover leads to climate cooling or other local climatic effects. Similarly, the dual effects of climate change and land degradation may have impacts on biodiversity that may further exacerbate land degradation, compromise the provision of ecosystem services and limit capacities to adapt to climate change. It is therefore important for research to focus on identifying ecosystems and areas where these feedbacks are most likely to occur, in order to identify options for climate mitigation and adaptation and move along a trajectory conducive to achieving LDN.

Assessing sensitivity to the interactions between climate change and land degradation is also in part a social science challenge. First, assessing the sensitivity of ecosystems and human populations to climate change and land degradation requires both scientific and locally held knowledge. By definition, land degradation must be assessed in relation to the objectives of those using the land, and locally held knowledge is usually necessary to appreciate the full effects of climate change on livelihoods and human well-being. Collecting and analyzing qualitative data from local communities can be time-consuming and expensive. However, the costs of doing this versus the costs of inaction need to be evaluated.

Second, in addition to considering the sensitivity of ecosystems to these processes, it is necessary to understand the sensitivity of livelihoods to the combined effects of climate change and land degradation. The sustainable livelihoods approach (Carney, 1998; Scoones, 1998) provides a framework for analyzing both the key components that make up livelihoods and the contextual factors that influence them, which may make a household or community more or less sensitive to the effects of a changing climate and land degradation (Eakin and Luers, 2006). Climate change and land degradation have the potential to disrupt established ecological and land use systems, which in turn may lead to the failure of food and water supplies, and the failure of livelihoods. This may in turn then limit the adaptive capacity of households when they are faced with other perturbations or stresses.

## 10.2 What can we do?

While accepting that further degradation is inevitable (e.g. due to pressures from growing populations and climate change) the concept of LDN emphasizes the restoration of degraded areas through an integrated landscape approach, to provide a range of ecosystem services (including climate regulation), whilst also protecting biodiversity and food, energy and water security. In the same way, no matter how successfully we mitigate climate change, further changes in our climate are inevitable. As such, it is essential to consider a range of response options.

There are a number of ways to enhance adaptive capacity and retain the integrity of ecosystems whilst maintaining sustainable livelihoods in the face of the interactive effects of climate change and land degradation. Adaptive responses need to balance the needs of the natural environmental, social needs (which vary with e.g. location, gender, age and socio-economic status), institutional needs (to facilitate cross-scale adaptations, establish incentives and shape behaviours), and the need to provide access to relevant information, technology and private sector engagement.

This is not always easy. Limited resources or distance to markets may mean that there are few alternative ways for some communities to pursue sustainable livelihoods. Other barriers are at higher levels – a lack of political will or power to facilitate adaptation, or poorly resourced or inflexible institutions that are slow to respond to change. In some cases there is simply a lack of information about adaptation options that are relevant to the challenges posed by climate change and land degradation. In other cases, the information and resources are available, but inaction arises from the high perceived risks associated with some adaptation options, especially if they challenge social norms.

Even when these barriers to adaptation have been overcome, some adaptations may be maladaptive, for example reducing the range of future adaptation options or increasing the vulnerability of other groups to climate change and land degradation. There is also a danger that one person's adaptation may contradict and nullify another's. It is therefore important to evaluate potential trade-offs between adaptations, so that complementary bundles of adaptations can be implemented together.

In this book, we have identified a number of ways in which communities in very different social-ecological systems around the world may be able to simultaneously adapt to climate change and land degradation. These include "climate smart" cropping and livestock systems that are adapted to future climates and protect the productive potential of the land. Ecosystem-based adaptations focus on enhancing the health and capacity of the natural environment to buffer change and enhance natural resilience to climate change and land degradation. Many ecosystem-based adaptations, such as those based on ecosystem restoration also restore and enhance biodiversity; a triple-win for climate change, land degradation and biodiversity.

In addition to this, there is a rapidly growing range of Sustainable Land Management (SLM) options that can enable land managers to adapt and be as resilient as possible in the face of a changing climate and avoid land degradation. In addition to facilitating adaptation, SLM has the potential to mitigate climate change, if practiced on a significant enough scale, by reducing or avoiding GHG emissions from degrading soils and biomass, whilst increasing rates of carbon sequestration in soil organic matter, which in turn enhances resilience to future land degradation and facilitates food production. Where SLM also promotes biodiversity, for example in agroforestry hedgerows, this too may be considered a "triple-win" for climate, land and wildlife.

Increasingly, the adaptations that are being developed in response to climate change and land degradation draw on multiple sources of knowledge. Rather than being derived from textbooks or based on scientific knowledge alone, adaptation options are increasingly drawing on locally held knowledge, often building on historic adaptations to climatic variability. By critically assessing and combining the best experiential "know-how" of local people with the more process-based "know-why" of researchers, it may be possible to develop future adaptations that are more effective, as well as better suited to the needs and preferences of the local people who need to use them.

There are already many methods available for monitoring changes in climate and the degradation status of land, but there has been less attention given to monitoring adaptation to these processes. Partly, this is a challenge of monitoring changes to ecological systems that will arise from the interactions between climate change and land degradation, which as we have discussed in this book, will be far from straightforward. Partly however, the challenge is to understand effects on social systems, including the capacity for different groups within society to maintain livelihoods, and identify ways to maintain the often overlooked cultural services that are provided by land. It is increasingly clear that there are many gaps in our understanding about links between climate change and land degradation (Box 10.1). To address many of these questions, it will be essential to pool knowledge from different sources to better understand the processes involved and the most relevant response options.

---

## BOX 10.1: REMAINING RESEARCH GAPS ON THE LINKS BETWEEN LAND DEGRADATION, DESERTIFICATION AND CLIMATE CHANGE

*Exposure and sensitivity to climate change and land degradation*

- How can we best characterize and understand the vulnerability and adaptive capacities of ecosystems (in particular agro-ecosystems) and human populations in affected regions, including regions newly susceptible to the consequences of climate change?
- What methodologies can capture the temporal and spatial dynamics of vulnerability and adaptive capacity? To what extent can temporal and spatial analogues be used to identify possible trajectories of vulnerability?
- How might the effects of climate change be moderated by interactions with other future social-ecological trends and drivers of change to make ecosystems and populations less vulnerable to land degradation?
- What trade-offs might exist between climate adaptation options in terms of their effects on ecosystem service provision and land degradation?
- Are there complementary bundles of adaptation options that can reduce trade-offs and create win-wins for both climate change and land degradation?

- At what spatial scale do vulnerability maps provide the most useful information to decision makers whilst at the same time retaining richness of information?
- What steps can be taken to deliver a more equitable distribution of adaptive capacity across different social-ecological systems?
- What preventative measures can be undertaken to prevent the erosion of adaptive capacity?
- How can climatic drivers prevent or speed up the land degradation and how can emerging opportunities be used to reach LDN in the context of a changing climate?

*Responses*

- How can we build efficiently on available knowledge, success stories and lessons learnt, to promote implementation of better adapted, knowledge-based practices and technologies?
- How do knowledge exchange activities, social relations and power shape the way knowledge is shared and created?
- What are the challenges associated with managing knowledge exchange at different organizational and spatial scales?
- How do contextual conditions (e.g. political, structural, legal and resourcing factors) and the way knowledge is understood and framed influence the way knowledge exchange strategies are developed within international policy programmes such as UNCCD and UNFCCC?
- What are the processes and mechanisms through which knowledge exchange activities (at different scales) generate beneficial outcomes for the ecosystems and human populations that are affected by climate change and land degradation?
- How do different research (disciplinary) and decision-making contexts influence the likelihood that knowledge exchange delivers beneficial outcomes for ecosystems and human populations?
- What formats should knowledge and information take to enable widespread sharing of success stories and upscaling across areas with comparable conditions?
- How can scientists and other stakeholders co-evaluate and jointly communicate success stories and adaptations?

*Monitoring and assessment*

- What are the most important variables in monitoring interactions and feedbacks between climate change and land degradation?
- How can we reconcile results from the monitoring of slow and fast variables?

- What resolution and frequency of monitoring provides optimal information to decision makers for important variables linked to climate change and land degradation?
- How can we identify the thresholds (temporal and spatial) at which adaptive practices and technologies may become maladaptive, such that their spread should be discouraged?
- How can we use modelling and mapping approaches to prioritize spatial areas for in-depth monitoring and assessment?
- What resources are needed and how do the costs of monitoring (action) fare against the costs of not monitoring (inaction) over short, medium and long time frames?
- How can we monitor and evaluate LDN at global, national and local levels?

## 10.3 Concluding remarks

We know that climate change is predicted to increase the number and length of droughts in many parts of the world, and growing global demand for food, particularly meat, is putting greater pressure than ever before on our land. This combination of climate change in a degraded landscape means that increasingly, in a severe drought, many farmers will experience substantial losses. This situation is not just theoretical: it is real and happening already in many parts of the world. If we do not find ways of tackling the *combined* challenges of land degradation and climate change, then many of the world's most vulnerable communities and ecosystems, particularly in drylands, may face both human and natural disaster.

It is often assumed that if we can just understand how the climate will change, we will know how communities and ecosystems will be affected, and we can think about how we might adapt. The reality is that climate change will interact with land degradation in quite different ways for different social groups in different places and over different time-scales, making it very difficult to predict what will actually happen.

If we want to understand how vulnerable a community or ecosystem is to the combined effects of climate change and land degradation, the first thing we need to know is how *exposed* the system is to these two processes. If the system is exposed to specific changes in climate and land degradation, then we can consider how *sensitive* the communities and ecosystems in that area might be to those changes. Will habitats still exist for key species? Will people still be able to earn a livelihood from that landscape in future, and will society still receive the services it relies upon from that place? If they are exposed and sensitive, we need to consider how easily they might be able to *adapt* to these changes, changing the way ecosystems and communities function so that species and livelihoods can still be maintained from that landscape. If the place we are looking at is not exposed or sensitive, or can adapt to these changes, then it is *resilient* in the face of climate change and land degradation. If it is exposed and sensitive, and cannot adapt, then it is *vulnerable* to these changes. The conceptual framework detailed in Figure 3.3 links each of these concepts, and is a powerful way of structuring the way we think about the effects of climate change and land degradation, so we can systematically assess and prioritize the likely threats, enabling decision-makers to take targeted action.

Using this framework, we have identified a number of key global vulnerabilities to the combined effects of climate change and land degradation (see Section 10.1). The most significant challenge in identifying these vulnerabilities is that these processes can in some cases be self-reinforcing, leading to feedbacks between climate change and land degradation, where each process exacerbates the other. The dangers posed by positive feedbacks are more immediate, and happening already, as the productivity of systems changes in response to land degradation processes that typically occur early on, like changes in species composition. The climate-cooling effects of a more reflective planet will however only happen once degradation processes have taken their full course. Although this may offer long-term hope, such hope is likely to come long after these regions have become uninhabitable and the degradation has become irreversible.

To reduce vulnerability, we have identified a number of approaches to adaptation that are to provide win–wins, tackling both climate change and land degradation at the same time. Some are "triple-win" options that can help support and enhance biodiversity and ecosystem services too, delivering goals for land degradation and poverty under the UN Convention to Combat Desertification, whilst tackling climate change and biodiversity loss, which are the focus of the other two Rio Conventions.

However, if we are putting adaptation measures in place, we need to know if they are working or not, so we can tweak them, to make sure they are heading for success. There are many methods available for monitoring biophysical changes. But if we want to identify and monitor the factors that increase or decrease vulnerability to the effects of climate change and land degradation, we need to monitor effects on livelihoods and well-being as well as ecosystem processes and services. Biophysical assessments will be necessary, but they need to be triangulated and interpreted in relation to socio-economic data within specific cultural settings

Finally therefore, we need far better co-operation and knowledge exchange both within the land management research and policy communities and between the land management and climate change communities. We need to mainstream participatory approaches so as to negotiate between diverse stakeholder perspectives in developing effective responses to climate change and land degradation. Some commentators point to the failures of participatory approaches that have been costly, time-consuming and had unintended negative consequences. Increasingly however, evidence supports key good practices in participation that are being shown to deliver results in a range of contrasting and often challenging contexts.

# References

Blaikie, P.M., Cannon, T., Davies, I., Wisner, B. 1994. *At Risk: Natural hazards, peoples vulnerability and disasters.* Routledge: London.

Bohle, H. 2001. Vulnerability and criticality: perspectives from social geography. *IHDP Update* 21: 3–5.

Carney, D. (ed.) 1998. *Sustainable Rural Livelihoods: What contribution can we make?* Department for International Development: London.

Eakin, H., Luers, H. 2006. Assessing the vulnerability of social–environmental systems. *Annual Review of Environment and Resources* 31: 365–394.

Hilhorst, D., Bankoff, G. 2004. Introduction: mapping vulnerability. In: *Bankoff, Vulnerability, Disasters, Development and People*, Frerks, G., Holhorst, T. (eds). Earthscan: London, pp. 1–24.

Reed, M.S., Buenemann, M., Atlhopheng, J., Akhtar-Schuster, M., Bachmann, F., Bastin, G., Bigas, H., Chanda, R., Dougill, A.J., Essahli, W., Evely, A.C., Fleskens, L., Geeson, N., Glass, J.H., Hessel, R., Holden, J., Ioris, A., Kruger, B., Liniger, H.P., Mphinyane, W., Nainggolan, D., Perkins, J., Raymond, C.M., Ritsema, C.J., Schwilch, G., Sebego, R., Seely, M., Stringer, L.C., Thomas, R., Twomlow, S., Verzandvoort, S. 2011. Cross-scale monitoring and assessment of land degradation and sustainable land management: a methodological framework for knowledge management. *Land Degradation & Development* 22: 261–271

Reed, M.S., Stringer, L.C., Dougill, A.J., Perkins, J.S., Atlhopheng, J.R., Mulale, K., Favretto, N. 2015. Reorienting land degradation towards sustainable land management: linking sustainable livelihoods with ecosystem services in rangeland systems. *Journal of Environmental Management* 151: 472–485.

Schwinning, S., Sala, O.E., Loik, M.E., Ehleringer, J.R. 2004. Thresholds, memory, and seasonality: understanding pulse dynamics in arid/semi-arid ecosystems. *Oecologia* 141: 191–193.

Scoones, I. 1998. Sustainable rural livelihoods: a framework for analysis. *IDS Working Paper 72.* Institute of Development Studies: Brighton.

Warren, A.S. 2002. Land degradation is contextual. *Land Degradation and Development* 13: 449–459.

# INDEX

A page reference in *italics* indicates a figure.